we're friends, right?

INSIDE KIDS' CULTURE

William A. Corsaro

Joseph Henry Press
Washington, D.C.

Joseph Henry Press • **500 Fifth Street, N.W.** • **Washington, D.C. 20001**

The Joseph Henry Press, an imprint of the National Academies Press, was created with the goal of making books on science, technology, and health more widely available to professionals and the public. Joseph Henry was one of the founders of the National Academy of Sciences and a leader in early American science.

Any opinions, findings, conclusions, or recommendations expressed in this volume are those of the author and do not necessarily reflect the views of the National Academy of Sciences or its affiliated institutions.

Library of Congress Cataloging-in-Publication Data

Corsaro, William A.
  "We're friends, right?" : inside kids' cultures / William A. Corsaro.
    p. cm.
  ISBN 0-309-08729-5
  1. Social interaction in children—United States. 2. Interpersonal relations in children—United States. 3. Children—Social networks—United States. 4. Social interaction in children—Italy. 5. Interpersonal relations in children—Italy. 6. Children—Social networks—Italy. I. Title: We are friends, right?. II. Title.
  HQ784.S56C67 2003
  303.3'2—dc21
                          2003009135

Credits:

Figures 3 and 4, photographs by Kathleen Barlow.

we're friends, right?

To my mother, Elizabeth Mahern Corsaro,
who inspired in me a love of all children
and a deep respect for childhood

About the Author

. . . .

William A. Corsaro is the Robert H. Shaffer Class of 1967 Endowed Professor of Sociology at Indiana University, Bloomington, and the world's leading authority on child ethnography. His ground-breaking research in the sociology of children, children's worlds and peer culture in a cross-cultural perspective, and early childhood education have revealed new and valuable insight into the unique world of children. He has devoted the past 29 years to extensive ethnographic fieldwork learning firsthand about children's culture and educational processes in preschools and elementary schools in the United States and Italy. He has also taught at the University of Bologna, Italy. His research has been featured on NPR, the BBC in London, and in the *New Yorker*. William Corsaro lives in Bloomington, Indiana.

Acknowledgments

· ·

Research reported on in this book was supported by the National Institute of Mental Health, the Spencer Foundation, the William T. Grant Foundation, and Indiana University, Bloomington. I wish to thank Suzanne Gluck for suggesting this project to me and for her help in finding my publisher. My editor at Joseph Henry Press, Jeffrey Robbins, has been consistently supportive, helpful, and wise. I wish to thank a number of colleagues and students who worked with me on the various research projects upon which this book is based. They are Hilary Aydt, Donna Eder, Jenny Cook-Gumperz, Francesca Emiliani, Ann-Carita Evaldsson, Laura Fingerson, Kathryn Hadley, David Heise, Douglas Maynard, Elizabeth Nelson, Thomas Rizzo, Katherine Brown Rosier, Heather Sugioka, and especially Luisa Molinari.

I wish to thank my wife, Vickie Renfrow, for her support and help and especially my daughter, Veronica Marie Corsaro, who read the manuscript and offered me an important perspective on the content and prose.

I am grateful to all the parents of the children I have studied over the years for giving me permission to share in their children's lives. I truly believe that preschool and elementary school teachers have the most important jobs in modern society. I have been fortunate to work with many outstanding teachers in the United States and Italy, and I am thankful for their insight, support, and friendship. Finally, I thank the many kids (some of whom are now adults) who let me enter their lives, become their friends, document their cultures, and tell their stories in this book. I hope I have captured at least some of the complexity of their childhoods and peer cultures.

Preface

· ·

Giving Voice to Kids' Culture

Kids are profoundly *social*. In my many years of observing in preschools, I have rarely seen a child grab a toy, book, or even a cookie and run off to play with or eat the item by himself or herself. Instead, the emotional satisfaction of sharing and doing things together is intense, especially when kids accomplish things together on their own without the aid or direction of adults. *Kids want to gain control of their lives and share that sense of control with each other.* In doing so they teach each other how to be social.

These two themes, control and sharing, were demonstrated in a wide range of activities in the childhood cultures of young children that I identified in my many years of observing in preschools in the United States and Italy. In this book I try to capture the lives of kids as they create and participate in the first of a series of peer cultures. In doing so I want to bring the voices of children to the adult debates about childhood. I want to convey what kids can tell us about how we can help them preserve and enrich their childhoods as well as prepare them for the adult world.

I am in a unique position to give voice to kids and their cultures because of my success in entering their worlds and the great breadth of my observations. Ethnographies of young children are rare, and most have been conducted in a limited range of space and time, usually in a single setting over a one-year period at most. I have carried out ethnographies of kids' cultures in preschool settings for nearly 30 years in the United States and Italy. During this period I have observed on a regular (often everyday) basis for a year or more in private middle-class preschools in Berkeley, California, and Bloomington, Indiana, in government-supported Head Start preschool programs for economically disadvantaged (mostly African-American) children in Indianapolis and Bloomington, Indiana, and in public preschools in Bologna and Modena, Italy. I have followed several groups of these children and continued my observations as they moved from preschools to elementary schools in Indianapolis and Bloomington and in Modena, Italy. I have also interviewed (both formally and informally) many of these children's teachers and parents about the children's educational experiences and peer cultures.

In examining kids' cultures I rely on a comparative perspective, often providing examples from upper-class and middle-class American kids, economically disadvantaged African-American kids, and primarily middle-class Italian kids. Kids from all three groups shared a number of features of peer culture, especially the desire to gain control over their lives and share that sense of control with each other. However, the kids from the three groups varied with respect to their interpersonal styles of interaction, the ways they formed and maintained friendships, and their strategies for engaging in and settling disputes and conflicts. These differences in many ways reflect the different experiences the children have with the adults in their lives and in their communities. The differences also reflect a diversity of concerns, interactive styles, and values in the kids' peer cultures that should be documented and appreciated. We should avoid the tendency to make quick judgments of such differences as deficient or even threatening to the dominant middle-class culture in the United States.

The comparisons involving Italian children, their preschools, and peer culture can teach us a great deal about the many advantages of investing more resources in the lives of young children in the United States. Italy has a long history of innovative social policy and government support of early education. It has paid off in that Italian preschoolers have developed rich peer cultures that are closely integrated into and enrich the everyday lives of their teachers, parents, and Italian society more generally than in the United States.

Finally, I decided to write this book to give voice to kids' cultures. I have learned a great deal from experiences with my many young friends. Over the years the kids I studied have made me a better researcher and, more importantly, a better person. Although it is my goal to share my experiences and knowledge with parents and all adults so that they can better understand and support children's childhoods, I also want to tell these kids' stories because they are important in their own right.

Contents

· ·

we're friends, right?

Introduction

The Importance and Autonomy of Kids' Culture

In the United States there is growing concern about a loss of childhood. We complain that we have less time to be with our children and that they grow up too fast. In many families both parents work outside the home and the number of working, single-parent families has increased dramatically. As a result, some children are placed in non-parental care during their first year of life and many others enter full-time day care and early education settings at age two or three. We are also alarmed by the growing power and extensiveness of the media. It seems that children are bombarded with adult images and information from an early age. Parents are uneasy about these changes. They worry that they might not be making the right decisions. So they often turn to experts for help.

Some of these experts bemoan the loss of innocence in contemporary childhood and advise parents to shield their children from the negative aspects of adult life. Others feel that children need more adult guidance, direction, and discipline. Often these experts argue that children are abandoned by their parents and turned over to non-parental care and the peer group. Thus, the argument goes, many young children lack traditional values and respect for authority and are prone to violence and instant gratification.

But what is it like to be a child in contemporary American society? What do *kids* have to say about the supposed erosion of their childhoods? The perspective and voice of the children are missing in this debate. The goal of this book is to provide a first-hand look at childhood from inside kids' cultures.

There are a number of reasons why kids themselves are not seen as important contributors to debates about their own lives and childhoods. First, many adults look to the future when they evaluate the state of childhood. They want their children to become healthy, happy, and productive adults, and they feel they are ultimately responsible for how their kids turn out. Many child-development experts encourage this way of thinking because they define and evaluate children by what they are going to be and not by what they currently are.

Parents surely affect the type of adult their children eventually become, but how much these effects go beyond genetic factors is hotly debated. In fact, in her book *The Nurture Assumption*, Judith Harris argues that, aside from supplying genes, parents have little to do with how their children turn out the way they do. Harris points to the importance of peers, arguing that kids aren't interested in becoming copies of their parents. Children want to be good at being children. Harris's critics argue that she overstates her case about the importance of peers over parents and in the process absolves parents of their responsibilities and gives them license to do a bad job of parenting. Although Harris draws attention to the importance of peers and children's culture, her main focus (like that of her critics) is still almost entirely on future outcomes—the type of adults children will become. Children and their cultures remain secondary in these debates.

While many adults look to the future when evaluating contemporary childhood, others reflect on the past. They reminisce about their own childhoods and believe that things have changed for the worse.

Thirty and forty years ago young children spent most of their time with other children—siblings and friends in their homes and neighborhoods. Mothers were often nearby to monitor their children's activities, fix them a snack, and put them down for a nap. Although these mothers played with their children and gave them attention, few felt the need to intervene frequently in their play or to devise structured learning environments. Most mothers believed that their young children surrounded by other children should enjoy and share their childhoods with each other.

Family structure has changed dramatically over the last 30 years. In 1970, approximately 30 percent of mothers with children under the age of six worked outside the home; the percentage rose to nearly 62 percent in 1999. Given this trend it is not surprising that young children are spending more of their time in preschools and kindergartens. In 1970, about 20 percent of all three- and four-year-olds attended private or public preschool; in 1999, that number had increased to more than 60 percent. There has been a similar increase in kindergarten attendance by five-year-old children (from 69 percent in 1970 to nearly 90 percent in 1999). Overall, approximately 38 per cent of all three- to five-year-olds attended preschool or kindergarten in 1970 compared to nearly 65 percent in 1999. Additionally, family size has decreased dramatically in the United States over the last 50 years. In the 1950s each American child had about 3.4 siblings; in the 1990s, the average number of siblings had fallen to 1.8. With more parents working, more children in child care, and fewer siblings, children are spending more and more time with peers outside the home.

But is the fact that children spend more time with peers such a bad thing and does it really differ that much from the past? Surely there is a need for more progressive family-leave policy in the United States and for higher quality and government-supported child care and early education. Still, as in the past, today's young children also spend most of their time with other children—among children who are seldom their siblings, but who are almost always their friends. The adults who are nearby are often mothers, but not the mothers of the children they care for. The overwhelming majority are often underpaid and underappreciated caretakers and teachers who love and dedicate their lives to children. They, too, monitor the children's activities, fix them snacks, and put them down for naps. Like the mothers of the past, these caretakers and teachers also believe that children should enjoy and share their childhoods with other children. And in high quality care and early education programs, teachers encourage and challenge kids and offer them opportunities to collectively create and participate in their own children's cultures.

When we focus too much on our children's futures by trying to make every aspect of their lives a learning experience or by brooding over the possible negative effects of every parental decision, we overly restrict their lives and steal away important moments of their childhoods. When we reflect nostalgically about our own childhoods and try to recreate the past in the lives of our children, we do our children a disservice because they must live their childhoods in the present. We have had our childhoods, and we cannot live our children's lives for them.

Does this mean that we should not try to influence our children or that we have no role in our children's lives outside the family? No, of course not. We can and should love, encourage, support, guide, and challenge our children. In our role as parents we have the duty to contribute in positive ways to our children's evolving membership in their culture—especially when they are young. However, we must realize that as parents we do not simply mold or shape our children. Children are active agents in their own socialization. In fact, kids creatively take information from the adult world to produce their own unique childhood cultures. In this sense, children are always participating in and are part of two cultures—adults' and kids'—and these cultures are intricately interwoven. Adults tend to overlook childhood cultures (especially those of young children) or view them as threatening (especially those of preadolescents and adolescents). I believe, on the other hand, that we can learn much from kids and that their cultures have an autonomy that makes them worthy of documentation and study in their own right. But to learn about kids' cultures from their perspective, we need to shed our adult point of view and get inside the children's worlds.

Getting inside children's worlds is difficult. We adults are bigger, more socially and cognitively mature, and more powerful. Children know and expect this of adults and it is not easy to overcome their preconceptions; to kids, no matter how hard an adult might try to act otherwise, he or she is still an adult. However, with care, patience, and persistence I have been able to overcome many of these barriers and get kids to see me as an atypical adult. More importantly, I have been successful at entering into their worlds as a special friend, to share and document their peer culture.

In my first study, once I was accepted by the kids, my goal was to demonstrate that they were active agents who contributed to their own socialization. However, I soon discovered that their worlds were much more complex than I had thought. It wasn't as I anticipated, that is, just that the kids were skilled social actors rather than passive agents. Little by little I began to see that I was not simply verifying young children's impressive social skills and the positive effects of peer interaction on their individual development. I found myself studying collective, communal, and cultural processes. I was documenting the children's creative production of and participation in a shared childhood culture. My full grasp of this revelation was gradual because I clung strongly to the typical adult tendency to try to interpret and evaluate almost everything children do as some sort of learning experience that prepared them for the future.

It is not that we adults fail to ever appreciate the special nature of children and childhood. We treasure children's spontaneity and joyfulness. We might even, at times, yearn to return to our own childhoods. Yet, though we might be aware of the ways kids differ from us, there is much that we do not grasp and understand about children's worlds. Further, we often distort and misinterpret what we don't understand by forcing it into our own adult perspective.

To gain insight into kids' worlds and peer cultures it is necessary to enter their worlds directly and be accepted as an atypical adult—a special friend who will not tell them what to do or attempt to control their behavior. Let me demonstrate how I accomplished such entry and acceptance in preschools in the United States and Italy.

1

"Yeah, You're Big Bill"

· ·

Entering Kids' Culture

I enter the outside play area of the preschool and walk up to two four-year-old girls, Betty and Jenny, who are sitting in the sandpile. As I get close to them, Betty says:

"You can't play with us!"

"Why not?" I ask.

"Cause you're too big," Betty replies.

"I'll sit down," I say as I plop down in the sand next to the girls.

"You're still too big," says Jenny.

"Yeah, you're Big Bill!" shouts Betty.

"Can I watch?" I ask.

"OK," says Jenny. "But don't touch nuthin'!"

"Yeah," says Betty. "You just watch, OK?"

"OK."

"OK, Big Bill?" asks Jenny.

"OK."

(Later Big Bill got to play.)

BECOMING AN ETHNOGRAPHER OF KIDS' CULTURES

Ethnography is the method anthropologists most often employ to study exotic cultures. The word "ethnography" is derived from "ethno" (people or culture) and "graphy" (the writing about or study of). You might have heard of the classic ethnographic study, *Coming of Age in Samoa*, by Margaret Mead. The ethnographic method demands that researchers enter, become accepted by, and participate in the lives of those they study. In this sense, ethnography involves "going native." As I noted in the preface, once I became convinced that children have their own cultures, I wanted to become part of those cultures and document them. To do this, I decided I had to enter into the children's everyday lives—to be one of the kids as best I could.

But how does a grown man go about being accepted into children's worlds? When I began my research, there were no established models to follow. So when I entered the first of many preschools I studied in the United States and Italy, I decided that the best way to become part of children's worlds was to "not act like a typical adult." In this chapter I describe how I went about doing this in several of the different early education settings I became part of and shared with kids, their teachers, and parents.

Let's start at the beginning, many years ago in Berkeley, California. I began in Berkeley because my dissertation director had a friend who agreed to sponsor my postdoctoral research and could also help me gain access to a preschool affiliated with the university.

BERKELEY, CALIFORNIA (1974-1975)—"A Big Kid"

In preparing for my research in Berkeley, I took the advice of one of the teachers and spent several weeks observing interactions in the school from a concealed observation area. The teacher, Margaret, told me that in the first weeks of school the children were still adjusting to the new setting, and parents and teachers were also a bit tense about the beginning of the new year. So she suggested that I observe from the one-way screened area that ran the length of the school's inside and outside. This viewing area was used by parents and for some observational research by developmental psychologists from a nearby university.

In my first days of observation, I was overwhelmed by the number, range, and complexity of the interactive events occurring before my eyes. On the first day, I had no clear idea of what to write in my field notes, so I just watched and tried to make general sense of things. In the following days, I began to focus on what happened when and where in the school and discovered a general routine. I made an inventory of the various activities in which the children participated, both those directed by the teachers and those they created themselves. I also gradually learned all the children's names and, to some extent, their various personalities.

During the third week I began to consider how I was going to enter into and be accepted by this group of kids who were becoming more familiar to me. Because I wanted to become involved directly in the kids' peer interactions, I knew that I did not want to be seen as a typical adult. The first step to discovering how to do this was to watch closely how the adults interacted with the children. Here is what I saw.

The adults were primarily active and controlling in their interaction with the kids. For example, parents and other adult visitors to the school often approached the kids, initiated interactions, and asked a lot of questions. Consider the following:

One day a visiting mother approaches a table where two girls are drawing. The mother watches for a while, bending over and looking down at the girls.

"What are you drawing?" she asks.

"A tree," one of the girls replies.

Now there's a silence as the girls continue their work.

"What color is the tree?" asks the mother.

"Green," says the girl, who does not look up but continues to draw.

"What else is green?" asks the mother.

Another silence and then the other girl says, "Grass."

The mother now straightens up, looks around the room, and moves off to another area.

Adults want to initiate conversations with children but are uncomfortable with the kids' minimal replies and their tolerance of what (to adults) seem to be long silences. Often, as in the example above, adults start asking test questions (things to which they already know the answer, like the color of a tree) to see what children are thinking about, or what they are doing, or simply to make the exchange a learning experience.

The teachers also asked a lot of questions, but they were more sophisticated in developing the learning potential of their conversations and interactions with the kids. They also directed and monitored the kids' play, helped in times of trouble, and told them what they could or could not do. Finally, adults (teachers or visitors) restricted their contact with the kids to specific areas of the preschool. Adults seldom entered the playhouses, outside sandpile, climbing bars, or climbing house.

Seeing how active and controlling adults were in their interactions with kids, I adopted a "reactive" entry strategy. In my first week in the school, I continually made myself available in child-dominated areas of the school and waited for the kids to react to me. For the first few days, the results were not encouraging. Beyond several smiles and a few puzzled stares, the kids pretty much ignored me. Of all the hundreds of hours I have observed in preschools, these were the most difficult. I wanted to say something (anything) to the kids, but I stuck with my strategy and remained silent.

On my fourth afternoon in the preschool, I stationed myself in the outside sandpile directly behind a group of five kids who were digging in the sand with shovels. They were doing "construction work" with four workers and a boss (four boys and a girl). The construction involved two of the boys digging a trench in the sand and another boy filling it with water while the fourth boy (the "dam stopper") stuck, pulled out, and restuck his shovel at various points in the trench to create a dam. He did this upon the orders of the girl boss. I watched this complex play for more than 40 minutes. Then the first two of the boys and, shortly after, the remaining two stuck their shovels in the sand and ran inside the school with the boss following them. I suspected that they did not plan to return and that the construction project was abandoned.

I was feeling ill at ease and considering my next move when I noticed Sue. She was standing alone near the sandpile about 20 feet away, and she was definitely watching me. I smiled and she smiled back, but then to my dismay she ran over near the sandbox and stood watching a group of three other girls. I then heard a disturbance near the climbing bars. I looked over to see that Peter had stolen (or so Daniel claimed) Daniel's truck. I noticed that a teacher had arrived to settle the dispute. When I looked back to the sandbox, Sue was gone.

I started to get up to go inside the school, but then I heard someone say, "What'ya doing?" Sue had approached from behind and was now standing next to me in the sandpile.

"Just watching," I said.

"What for?" she asked.

"Cause I like to."

Then Sue asked my name. I said (and this turned out to be important), "I'm Bill and you're Sue."

Sue took two steps back and demanded, "How do you know my name?"

I now did something adults seldom do when talking to young children, especially if they think kids will not understand the answer. I told the truth with no attempt to simplify.

"I heard Laura and some other kids call you Sue."

"But how do *you* know *my* name?" Sue asked again.

Sticking to my guns, I repeated that I had heard other kids call her Sue. She gave me a puzzled look, twirled around, and ran into the school.

So here I was. After spending several days trying to become one of the kids, finally a child talks to me and I scare her off! But then Sue re-emerged from the school and came running back to me with Jonathan by her side.

When they reached me, Jonathan asked, "What's my name?"

"Jonathan," I replied.

"How do you know my name?"

"I heard Peter [one of his frequent playmates] and some other kids call you Jonathan," I said.

"See, I told you he knows magic," said Sue.

"No, no, wait a minute," cautioned Jonathan. "Who're those kids over there?" he asked, pointing to Lanny and Frank.

"Lanny and Frank," I responded confidently. I knew all the kids.

Jonathan looked around, trying to find a hard kid and he then asked me to name three more. I identified them all easily.

With a sly smile Jonathan then asked, "OK, what's my little sister's name?"

Jonathan thought he had me. But I actually knew his sister's name. The secretary at the school had provided me with a roster that listed the names of the children, their parents, and their siblings. I memorized much of this information and, fortunately for me, I remembered Jonathan's sister's name.

"Alicia!" I declared.   I was feeling good now.

Jonathan was very impressed. He looked at Sue and said, "I can't figure this guy out." He then ran off to tell Peter and Daniel.

Meanwhile, Sue handed me a shovel.

"You wanna dig?"

"Sure," I said, taking the shovel.

We shoveled sand into the buckets and soon we were joined by Jonathan, Peter, and Daniel. Peter and Daniel asked me if I knew their names. I did, of course, and told them. Then we all started to shovel and the kids organized another construction project and I was assigned the role of worker. Christopher and Antoinette also joined us and the play continued for 20 minutes or so until one of the teachers announced "clean-up time," whereupon we reluctantly put away our shovels and went inside for meeting time.

For several days after this breakthrough, the children began to react to my presence (ask who I was) and invite me into their play. Although I was able to observe, and in many cases participate to some degree in the kids' play, their acceptance of me was gradual. During the first month, the kids were curious about me and why I was around every day. They asked lots of questions that followed a general sequence: "Who are you?" "Are you a teacher?" "Are you gonna play a game with us?" (that is, ask them to be in one of the research experiments that occurred routinely in this lab school) "Are you a Daddy?" and "Do you have any sisters or brothers?" The pattern here is important. The children moved from general questions about adult characteristics to the last question about siblings, which is one kids typically ask of each other.

At the time of this first study, my answer to all the adult information questions was "No" because I was not a teacher, experimental researcher, or a father. But I do have siblings—seven of them! My having so many brothers and sisters piqued the kids' curiosity about me. However, they were hesitant to believe me, and some asked, "For real?" Then to their delight I named them all. Being from a big family helped solidify my acceptance and standing in the group.

I am not claiming that the kids quickly accepted me as one of them. I have not in all my many years in preschools ever been seen totally as one of the kids. Even in Italy where I was seen as an adult incompetent because of my limited knowledge of Italian, I was still an adult. I am just too big to be a kid. Thus, the nickname that surfaced near the end of the first month at Berkeley in the scene I described earlier is important. I became accepted as a different or atypical adult—a sort of big kid.

My status as a "big kid" was demonstrated in a number of ways in my initial ethnography. First, I was allowed to enter ongoing peer activities with little or no disruption. I could move into the playhouses, sandpile, and even climbing structure without much comment beyond a few smiles and some laughter. Second, I had little or no authority when compared to other adults. Given my desire to be part of the kids' culture, I refrained from controlling their behavior. However, on those few occasions when I feared for their physical safety my "Be careful" warnings were always countered with "You're not a teacher!" or "You can't tell us what to do!" Finally, throughout the school year, the kids demanded that I be a part of the more formal peer activities. At birthday parties, for example, the kids insisted that I sit with them (in a circle) rather than on the periphery with the teachers and parents. Also, several of the kids demanded that their mothers write my name, along with those of their playmates, on cookies, cupcakes, and valentines that were brought to school on special days.

Before leaving the Berkeley part of my story, I should note that as an atypical adult I came to have a special relationship with the kids, but this relationship varied from child to child. In all the settings certain kids became special friends. In Berkeley it was Martin. Martin took to me early on and often looked for me when deciding on a group to enter in free play. Noticing that Martin was becoming a bit too dependent on me, I often slipped away from certain play activities once Martin got involved. Soon I discovered that Martin was fine on his own, but he still considered me one of his best buddies.

One day this became very clear to me when his mother stayed around after bringing him to school to talk with one of the teachers.

"Which little boy is Bill?" she asked.

"We don't have a Bill," responded Margaret. "Except for Bill Corsaro, but he is here doing research."

"Oh, I remember now signing a consent from a William Corsaro," said Martin's mother. "But Martin talks about Bill all the time and a book he has, so I thought it was another boy in the class. Martin keeps asking if he can have a book like Bill's to bring to school."

The book Martin's mother was referring to was the small notebook I always carried in my back pocket. After observing an episode of peer interaction, I often slipped away to a secluded area of the school and jotted down a few notes to be expanded later that evening. Martin asked me about the notebook once and I told him that I liked to write things in it to remember what happened. He sort of shrugged at this explanation and did not mention the notebook again.

So when I talked with his mother that day, I explained all this and offered to bring a notebook for Martin the next day. He was all smiles when I gave it to him and helped him put it in the back pocket of his jeans. It was a snug fit in the small pocket, but Martin did not take it out once it was inside. He patted his pocket now and then throughout that first day and brought the notebook most days to keep in his pocket. In this way he could be like Bill, a sort of junior ethnographer!

BOLOGNA, ITALY (1983-1986)—"An Incompetent Adult"

I was apprehensive about field entry in the first Italian preschool I studied because of my limited abilities in conversational Italian at that time. As it turned out, this apprehension was shortlived. With the help of Italian colleagues I gained entry to a preschool and presented my research aims (basically, what was it like to be a child in the school) to the teachers. In Italy, preschool is government funded and more than

96 percent of Italian three- to five-year-olds attend before entering the first grade of elementary school at six years of age. The school I became a part of had 5 teachers and 35 children in a mixed age group of three- to five-year-olds.

On my first day at the preschool, the teachers introduced me to the kids as someone from the United States who would be coming to the school to be with them throughout the year. Relying on the "reactive" strategy of field entry I had first used in Berkeley, I entered play areas, sat down, and waited for the kids to react to me. It didn't take long. They began asking me questions, drew me into their play activities, and over time defined me as an atypical adult.

Somewhat to my surprise my acceptance by the Italian children was much easier and quicker than it had been by the American children. For the Italian kids as soon as I spoke in my fractured Italian I was unusual, funny, and fascinating. I was not just an atypical adult, but also an incompetent adult—not just a big kid but sort of a big dumb kid.

The first thing they noticed was my accent, but they quickly got used to it and then realized that I often used the wrong words (bad grammar) and more often than not made little sense (bad semantics). At first they had fun laughing at and mocking my mispronunciations. But soon they became little teachers, correcting my accent and grammar and even repeating and adjusting their own speech when I couldn't understand them. At times they acted out words and frequently consulted in small groups, often laughingly calling to others, "Guess what Bill said now!" Before long we were doing pretty well and my confidence in communicating with the kids grew. I specifically remember one small triumph.

I was sitting on the floor with two boys (Felice and Roberto) and we were racing some toy cars around in circles. Felice was talking about an Italian race car driver as we played, but because he was talking so fast I could understand only part of what he was saying. At one point, however, he raced his car into a wall and it flipped over. Then I clearly heard him say "*Lui è morto,*" and I knew this meant, "He's dead." I guessed that Felice must be recounting a tragic accident in some past Grand Prix event. At that moment I remembered and used a phrase that I had learned in my first Italian course: "*Che peccato!*" ("What a pity!"). Looking up in amazement, Felice said: "Bill! Bill! *Ha ragione! Bravo Bill!*" ("Bill! Bill! He's right! Way to go Bill!"). "*Bravo Bill!*" Roberto chimed in.

Then Felice called out to other children in the school. Several of the kids came over and listened attentively as Felice repeated the whole story of the tragic accident and then added: "And Bill said, '*Che peccato!*'" The small group cheered and some even clapped at this news. Not in the least embarrassed by all the attention, I *felt good— like one of the group*! I was no longer an adult trying to learn about kids' culture. I was in. I was doing it. I was part of the action!

Things were not going as well with the teachers. In fact, confusion and communication breakdowns were frequent during my first months in the school. There were a number of reasons for these problems. First, the teachers and I were self-conscious about these language problems. For the teachers, it was because they knew only one language and for me it was because my Italian was poor. Second, we tried to talk about rather abstract topics (like early education policy in the United States) in contrast to the more here-and-now conversations I had with the kids during their play. Third, the teachers were not as good at adjusting their speech as the kids were. They would start off talking slowly and were careful to avoid difficult constructions and idiomatic expressions. However, after a conversation was under way, things sped up, complicated phrases emerged, and I got confused. When I expressed confusion, the teachers often got a bit flustered and insisted we start over, and as a result, we seldom got very far in these early attempts.

Given our difficulties, the teachers were surprised by my apparent communicative successes with the kids. On several occasions I saw one or another of the teachers call children over to ask them what we had been talking about. The kids had no problem telling the teachers what they and I had said. These explanations prompted the teachers to ask me why I could communicate so well with the kids. I told them that the children and I talked about simpler and more direct things related to the kids' play. While still a bit perplexed, the teachers accepted this explanation, and over time as my Italian improved, so did my communications with the teachers.

Importantly, however, the children's discovery of my communicative problems with the teachers was a special aspect of our relationship. They could talk with me and I with them with little difficulty, but it was apparent to them that my communication with the teachers was not as easy. In fact, several parents told me that their sons or daughters came home and told them: "There is this American, Bill, at the school and we can talk to him, but the teachers can't!" In short, the children saw my relationship with them as a partial breakdown of the control of the teachers.

The nature of my special relationship with the kids was nicely captured in a school project. Early on in the school year, all the children drew small self-portraits on separate sheets of paper. These individual portraits were then placed in a much larger group picture with the title: "*INSIEME DELLE FACCE DEI BIMBI DELLA DUE TORRE*" ("ALL TOGETHER THE FACES OF TWO TOWERS"). The large picture was displayed on the wall of the main meeting room of the school. *Due Torre* was the name of the school and the larger picture captured the communal nature of the school's curriculum. We can see the picture of the self-portraits in Figure 1.

Later the teachers asked the children to say a little about themselves. The teachers recorded their responses, typed them up, and placed them in a portfolio for each child along with the class portraits and other materials produced over the course of the year. In describ-

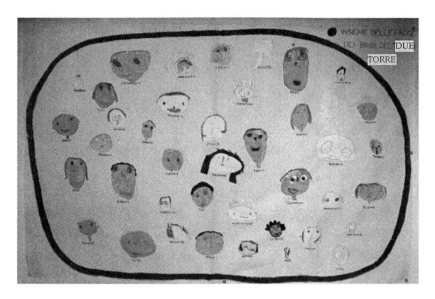

FIGURE 1 The children of Due Torre.

ing themselves, most kids referred to physical features, said that they had brothers or sisters, pets, what they liked to do, and so on. However, one girl, Carla, had only one simple response: *"Avevo una borsa."* ("I used to have a purse."). Despite urgings from the teachers and her classmates, Carla would say no more. I assumed the lost purse was awfully important to her.

After the kids finished their self-portraits, the older ones had the privilege of drawing portraits of the adults. This group included the teachers, the *dade* (women who worked in the school serving food and cleaning, but also at times acting as surrogate grandmothers for the kids), and me. These pictures were also placed into a larger group portrait and displayed alongside the children's group portrait with the title: *"INSIEME DEGLI ADULTI DELLA DUE TORRE"* ("ALL TOGETHER THE ADULTS OF TWO TOWERS"). It is not hard to recognize me in this group, shown in Figure 2.

FIGURE 2 The adults of Due Torre.

After the children had said something about themselves, they were given the opportunity in a group meeting to offer comments and descriptions of the adults. The children described the physical features of the teachers and *dade* and also offered some comments on their personalities. The kids said that some of the teachers were nice, but also a bit severe and raised their voices when the children misbehaved. Now we have arrived at the main point of this narrative of the drawings and descriptions. Here is what the children said about me.

Bill è un uomo alto e giovane. Ha i capelli neri, gli occhi marroni e porta gli occhiali, ha la barba. Viene sempre a scuola e gioca con i bimbi, è buono. Bill è Americano e non italiano, si capisce dalla lingua. Con i bimbi parla in italiano: è bravo.

(Bill is a tall, young man. He has black hair, brown eyes and wears glasses, he has a beard. He always comes to school and plays with the kids, he's good. Bill is American and not Italian, he understands the language. With the kids he speaks Italian very well.)

The children's own description captures very well their perceptions of and feelings about me. In their eyes I am a tall, young man (while in reality my height is just below average for American males) and I am good because I always come to school and play with them. In this way I am seen as a friend. Further, the relationship is special because even though I am an American and not an Italian I understand the language and *with them I speak the language very well.*

Despite these kind words about my language ability, the children never tired of teasing me about my mistakes when speaking or my failures to understand something someone had said. The youngest children most enjoyed such teasing. In fact, the kids often extended my incompetence in language to other areas of social and cultural knowledge.

Once we made a field trip to a zoo and theme park that had scale models of dinosaurs. During our visit I pointed out to a small group of kids (in very good Italian, I might add) that the particular dinosaur we were looking at had lived in the same place that I did in the United States. In fact, I knew I was correct about this because the sign with the exhibit said as much. The kids laughed uproariously at my comment. One boy, Romano, shouted out, "Bill, he's crazy! He says the dinosaur lived in the United States." Then pointing to the dinosaur he added, "But you can see it lived right here!" Given the logic of that rebuke, I made no attempt to protest the criticism of my comment.

My work in Bologna was the first time that I returned to a preschool for a second year. The three- and four-year-olds were a year older when I returned in May 1985. The anticipation of my return had been piqued by an exchange of letters with the children and teachers. I was greeted on arrival by the children and the teachers, who presented me with a large poster on which they had drawn my image and printed: "*Ben Tornato, Bill!*" ("Welcome back, Bill!"). After handing me the poster, the kids swarmed around me, pulled me down to my knees and each child took a turn embracing and kissing me. In the midst of the jubilation I noticed a few new faces—three-year-olds who had entered the school during my absence. One or two of these little ones shyly came up to touch me or to receive a kiss.

Later in the day after the commotion had settled, I was sitting at a table with several children who were playing a board game. I noticed a small boy, whose name, I later learned, was Alberto, eyeing me from a distance. He finally came over and asked: "Are you really Bill?" "Yes, I'm really Bill," I responded in Italian. Alberto, smiling, looked me over for a few seconds and then ran off to play with some other children.

One important aspect of this vignette for our discussion is its relation to my participant status in the local peer and school cultures. The children's jubilant marking of my return to the school was certainly related to the length of my absence—absence does indeed make the heart grow fonder. However, the closeness of my relationship with the kids went well beyond the joy accompanying the return of an old friend. Several ethnographers of children have pointed to the importance of developing a participant status as an atypical, less powerful adult in research with young children. In this case, as I argued earlier, my very foreignness was central to my participant status. My limited competence in the Italian language and lack of knowledge of the workings of the school led the children to see me as an "incompetent adult" whom they could take under their wings to show the ropes.

A second important aspect of the story is its capturing of the importance of longitudinal ethnography when studying young children. Recent theoretical work in this area is critical of traditional theories of socialization and child development for their marginalization of children. Traditional views focus on individual development and see the child as incomplete—in the process of movement from immaturity to adult competence. The new approaches eschew the individualistic bias of traditional theories and stress the importance of collective action and social structure. Longitudinal ethnography is an ideal method for such a theoretical approach, particularly when it aims to document children's evolving membership in their culture and when focused on key transition periods in children's lives. My return to the school was my first attempt to extend the longitudinal design of my research toward this ideal.

Let's return to our story to consider the richness of longitudinal ethnography. I did not simply return to my field site and renew my research. Traces of my continued presence were sketched by the children and teachers in their reflective talk about their past experiences with me. The memories and emotions evoked by these informally occasioned discourses were deepened and intensified by a series of more focused activities: their reading and discussing of letters and cards I sent them; their construction and enjoyment of a gift from me (a Halloween mobile of swaying jack-o'-lanterns, witches, spiders, and ghosts, along with a description of the wondrous but foreign children's holiday symbolized in the mobile); their composition of letters and art work to send to me; their discussion and anticipation of my return; and their construction of the poster to commemorate the homecoming. A version of these discourses and activities was also produced in my world—in discussions in my family, with my colleagues, with my students, and in my research reports.

Thus, the homecoming did not mark the beginning of a new phase of a longitudinal study, but rather a continuing evolution of my membership in this group. In turn, the documentation of and reflection on this evolution are of central theoretical importance for grasping both cognitively and emotionally the nature of the children's evolving membership in the local peer and school cultures of this educational institution.

Finally, there is the ending of my story and the young boy, Alberto. In his interactions with his peers and teachers over the course of his first year, this mysterious Bill had become somewhat of a legend to Alberto. Thus, Alberto, being rather a doubting Thomas, desired direct confirmation of my status. Alberto's interest in and fascination with me illustrates how the participant status of the ethnographer becomes embedded in the network of personal relations of those he studies over time in longitudinal research. Although Alberto needed to confirm the reality of my existence, he was very much influenced by what he had learned about me from the other kids. For example, he quickly seized on and relished my status as an incompetent adult.

A few days after my return, several kids were telling me about something that happened during my absence. The story had to be halted and repeated several times because I had trouble understanding. During the last retelling, Alberto joined the group and threw up his hands laughing: *"Ma uffa! Bill. Lui non capisce niente!"* ("Oh brother! Bill. He doesn't understand anything!"). It becomes somewhat easier for an adult to empathize with the lower status of children in society when he finds himself the butt of successful teasing of this sort by a three-year-old.

INDIANAPOLIS HEAD START (1989-1990)—"A New Friend"

Indianapolis is my hometown and when I met with the director of the Head Start center and the teachers I was to work with, we found that we shared many experiences growing up in the city. They were quick to accept me into the center. However, when I told them of my plans to visit the center twice a week over the school year to learn about the children's peer interaction and culture, one teacher was doubtful. "What do you want to do that for?" she asked. She was convinced that I would soon become bored or quickly find out all I needed to know. But after I had stayed true to my word for three weeks, the teachers began to look forward to my visits and we established a good rapport.

Things also went well with the children, who quickly pulled me into their play activities. However, my early experiences in the Head Start center were in one way completely novel for me. What was different about the Head Start study was that I was a white man in a world of mainly black women and children. For the first time in my life I was spending considerable time in a setting where I was a minority, as all but one of the teachers and staff in the center and the overwhelming majority of the children were African-American. Although I was very conscious of this fact, the kids seemed unconcerned. Over the first couple of weeks, several of the children asked me if I was Brandon's (a Latino boy's) father. I said I was not and that I was at the school to be and play with them. About two months into the study, one girl, Tamera, came up to me and said: "Bill, you're white!" Not knowing exactly what to say, I replied, "Yeah, I am." And that was that.

During the third week in the school something important for both the kids and the teachers happened with regard to my acceptance and participant status in the school. The Head Start center was located in an old elementary school and, unlike most preschools, there were no bathrooms for the children in the classrooms. Preschool children frequently need to use toilets and are too young to be allowed to travel to bathrooms located outside the classroom on their own.

Therefore, one of the teachers had to take the children as a group to centrally located bathrooms twice each session. I went along on these trips and watched as the teacher lined the children up along the wall outside the bathrooms. She then sent three or four boys into the boys' room and the same number of girls to the girls' room. She waited a few minutes, entered each bathroom, and hurried the children along, and then sent in the next group, until all the children had a turn. We then walked back to the classroom with the teacher reminding the children to stay in line, walk slowly, and be quiet so as not to disturb other classes.

I could tell that this was not a pleasant chore for the teachers. However, I was still surprised when one day in the morning group, a teacher asked me to take the children to the bathroom. This request seemed perfectly reasonable. After all, it was not a difficult task. Besides, if I was going to tag along when the teachers took the kids, why couldn't I take them down myself?

The problem was that I did not want to be seen as an authority figure by the children, and I had talked with the teachers about this aspect of my research. However, it was clear they did not think this small chore would cause me problems or they just didn't make the connection when they asked for the favor. I decided that it was best to agree to help out and hoped it would not be too much of a challenge to my relationship with the kids. It turned out I got much more than I bargained for, at least on our first trip to the bathrooms.

Things started out fine. I noticed a few smiles on the kids' faces when the teacher said I would be taking them. They were told to be on their best behavior as we exited the room. In the hallway and down the stairs they were like little angels. There was no talking or running; even the line was perfectly straight. They also were orderly as they lined up against the wall (boys near the boys' bathroom and girls near the girls').

I sent the first four boys in line (Charles, Luke, Joseph, and Antwaan) into the bathroom and also sent in four girls (Cymira, Tasha,

Michelle, and Lamecca). After a few minutes I heard a lot of noise in the boys' bathroom.

"What are they doing in there?" asked Jeremiah. This was the same question I was asking myself. When I went in to find out, I immediately knew I was in trouble. Joseph had wadded up several paper towels and was throwing them at the other three boys. Antwaan was standing at the sink with the cold water on full blast while he flung his hand under and then upward to spray water around the room. Meanwhile Charles and Luke were laughing loudly as they stood at an angle over the urinals trying to pee over each other's streams into the adjoining urinals.

"Hey, you guys," I said. "Cut that out and come back outside."

"You can't tell us what to do," said Charles who had at least straightened around and was peeing in his own urinal.

"Yeah, he's right," added Antwaan. "You're not a teacher."

Now hearing a lot of noise outside, I ran back out there. All the children wanted to have their turns and asked me when they could go in. Brandon was the most insistent, moaning "I gotta pee!" I had to go myself, but that was the least of my worries. I went back in with the boys and realized that trying to be stern was not going to help. Charles and Luke had now joined Joseph in throwing the paper towels, one of which hit me in the back of the head as I stopped Antwaan from throwing water by shutting off the faucet.

Before the boys could challenge me, I said: "I'm not a teacher, but Mrs. Green's class will be coming soon. If you guys don't get back outside, *we'll all get in trouble.*"

"Yeah, Bill's right," said Charles. "We better go back out."

The other boys agreed and I quickly ushered in the remaining boys, including Brandon, who raced in as fast as he could go. Thank God he had not wet his pants. Now for the first time I realized that the first four girls had still not come out, and there was a lot of noise coming from the girls' bathroom. I ducked my head in, but Tasha shouted out

"No boys allowed!" The teachers didn't have this problem as they entered the boys' room to hurry them along without concern. I decided to accept Tasha's warning. However, I was prepared to deal with the situation now.

"I think Mrs. Green and her class are coming," I said loudly.

"Uh-oh," I heard Michelle exclaim.

"Yeah, let's go," said Cymira. And soon all four of the girls were out and the rest of the girls were in.

The second shift of children played around a bit but were quick to heed my warning about Mrs. Green's class. Soon all the children were finished and we were lined up and ready to go. Several of the kids were smiling, and Charles said, "It's fun to go to the bathroom with Bill!" Now we started back to the room and the children were as well behaved as they were on the way down.

Back in the room, Mrs. Jones said, "You took a while. You better have not given Bill any trouble."

"We didn't," replied Charles looking at me with a smile.

"We like going with Bill," added Tasha.

I felt safe. I had cleaned up all the paper towels. The floor was still pretty wet in the boys' bathroom, but it would probably be dry by the time Mrs. Green's class got there.

After a few days the word spread to the afternoon class about my bathroom responsibilities and I was asked to take charge of bathroom time in that class as well. The kids gave me a hard time on the first trip, but now I was better prepared. Actually, this role brought me closer to the kids because they always knew they could play around a bit on these bathroom trips and could give me a bit of a hard time. Still, they realized that there was a limit to their horseplay. As was the case with the Italian children, we had certain experiences we shared away from the control of the teachers. Thus, my status as a special and fun adult was solidified.

Over time I became more and more involved in the children's activities in the school. The Head Start kids liked to tease each other and engage in what the anthropologist Marjorie Goodwin has called "oppositional talk." This type of competitive teasing and joking was rarely taken as offensive by the children. In fact, clever oppositions or retorts were often marked with appreciative laughter and comments like "Good one" or "You sure told her, girl!"

After several months I grew accustomed to receiving verbal jabs from the kids and on a few occasions returned a few of my own. Once, Charles noticed some young adult males from the local community who had entered the classroom to help the teachers prepare for an upcoming festival. I was eating lunch at a table with Charles and several other kids when Charles asked: "Are they gonna rap at the festival?"

"Yeah," I responded, "they're gonna rap you on your head!"

All the children laughed loudly, including Charles, who said, "Good one, Bill. Good one."

Near the end of the year at Head Start, I began to do some videotaping. Like I have done in all settings, I videotaped near the end of my observational period and had an assistant do the actual taping. In this instance my student, Katherine Rosier, came to do the taping and later was to carry out an intensive interview study of the children's parents. When Katy and I came into the room, Cymira ran up to us and asked, "Bill, is she your mother?" I responded that Katy was too young to be my mother. I said she was my friend and was going to help me make a videotape of the kids. Given that Katy was a number of years younger than me, it would seem to be obvious that she was not my mother (she sure saw it that way). But by this time the kids had accepted me as part of their group and when someone came to school with them it was usually a parent.

MODENA, ITALY (1996-2001)—"A Newcomer"

In Modena, Italy, I carried out a study of children's transition from preschool to elementary school with my Italian colleague, Luisa Molinari. We continued our study through observations and interviews throughout the children's five years of elementary school. The main focus of the study was on the children's last five months in preschool and their first four months in first grade.

My first days in the Modena preschool were a new challenge for me. For the first time I was in a preschool where I was the only true novice. In previous research, I had entered schools at the start of the term and at least some (if not all) of the kids were, like me, new to the setting. Furthermore, in this instance not only was I entering the group at the midpoint of the school year, but almost all of the children and teachers had already been together for two and a half years. This fact, coupled with my foreignness, led many of the kids and adults to be very curious about me during my first days in the school.

As I had done in past research, I moved into play areas, sat down, and let the kids react to me. Several of the older and more active kids in the group (Luciano, Elisa, and Marina) often told me what was happening and generally took charge of me during the first few weeks. They escorted me to the music and English classes, and I overheard them making reference to my presence to children in the other five-year-old and the four-year-old classes in the school and reporting that "Bill is part of our class!"

Even though the kids liked the idea of having me in their class, they, like the Bologna kids, made fun of my mispronunciations and bad grammar, often claiming that "*hanno capito niente*" ("they understood nothing") when I talked. Also, several children often patted my stomach, laughing about my "*pancia grande*" ("big belly"). One day, after I had been observing in the school for three weeks, I was sitting in an area where a girl, Carlotta (who frequently teased me), and several other girls were playing with some dolls. Carlotta suddenly pulled up my sweater, stuck a doll in, and called out to everyone, "Look, Bill's pregnant!" She then pulled the doll out to roars of laughter from the other kids.

The children were also quick to dismiss some of my ideas or claims. Once when playing in the outside yard with several kids, I noticed Dario, Renato, and Valerio gather some sticks and place them on the ground under the climbing bars. They protected their sticks from the others and there was some discussion of fire. So, I mentioned that Indians start fires by rubbing sticks together. Renato and Valerio decided to try this, but Dario said (in so many words) "Bill's *'pazzo'* ('crazy'), he doesn't know what he's talking about, and it won't work." The others quickly agreed and instead used the sticks to stir leaves.

On the other hand, the kids realized that, as an adult, I did have certain skills that were useful to them. Once Renato, Angelo, Mario, and Dario were playing with plastic grooved building materials. They handed me some pieces that were stuck together and asked if I could get them apart. I accepted this task willingly, but soon realized that the pieces were stuck much tighter than I thought. In fact, I first pushed with all my might, to no avail. One of the teachers, Giovanna, walked by, laughed, and said that the children had found a practical use for me. I now guessed that many of the pieces had probably been stuck together for a long time. Just as I was about to give up, I tried holding one piece on the edge of the table with the other hanging over the edge. I pushed hard and the pieces popped apart. Angelo and Renato yelled: "Bravo Bill!" and immediately handed me several more pieces. I easily separated the first two with my inventive method, but then I ran into trouble again as several pieces just would not budge. Meanwhile the boys were copying my method with some success, so I kept at it. I then noticed that Angelo and Mario were gathering up all the separated pieces and putting them back in the box. They told several other children that Bill got them apart, but they were not to play with them. I wondered about this. Were they afraid that pieces would just get stuck back together again? In any case I continued working on the unpleasant task until, to my relief, I heard Giovanna say it was time to clean up the room.

One morning after I had been observing in the school for about five weeks, Giovanna was reading a chapter of the *Wizard of Oz* to the children. After about 10 minutes of reading and discussion, she was called away to take a phone call. As she left she handed me the book, suggesting that I continue reading the story. Aware that it would be a difficult task for me, the kids yelled and clapped, thinking that this was a great idea. I immediately had a problem pronouncing the word for "scarecrow" which in Italian is "*spaventapasseri.*" The kids laughed and hooted at my stumbling over this and other words. Some even fell from their seats in pretend hysterics at my predicament. My task was made even harder because there seemed to be a "scarecrow" in every other sentence. To my relief Giovanna returned and, when asked how I did, the kids laughed and said I could not read well. Sandra yelled out, "We didn't understand anything!" Giovanna then took the book back from me, but the kids shouted: "No, we want Bill to read more!" Taking the book back, I struggled through another page amidst animated laughter from the children and handed the book back to Giovanna saying, "That's enough for now."

There are two aspects of the children's response to my problems with the language that were different from my earlier experiences in Bologna. First, in Bologna I observed a large, mixed-age group where there was wide diversity in the children's literacy skills. Also, although the Bolognese children were introduced to reading and writing, it was not a central part of the curriculum. In this group of five-year-olds in Modena, lessons and activities related to reading and writing were now everyday occurrences in these last months of their final year in the preschool. Although they laughed at my errors, they knew I could read, and they identified with my language problems to some degree. Second, the children in Modena were also studying English and they realized that I was competent in this foreign language that was very difficult for them. In short, it was reassuring to them that this new adult in their midst shared some of their same experiences and challenges.

Language was a central aspect in my acceptance by both the kids and teachers. My Italian had improved considerably since my earlier work in Bologna. I could converse easily with the teachers in the Modena classroom. Still, the teachers (Carla and Giovanna) realized that I was far from fluent in Italian and liked to tease me about it.

In one learning activity, the children were shown several common household objects that were then put into a bag. The teachers asked each child to reach into the bag and, without looking, touch, handle, and identify the object they selected, and then pull it from the bag. After each child had a turn, Carla asked me to reach into the bag. She knew, of course, that I could easily identify the objects, but she also suspected that I might not know the Italian names for several of them. I got hold of a can opener and immediately realized I was in trouble. I stuttered a little and then said in Italian, "It's a thing to open things." Carla and Giovanna laughed loudly and one child, Sandra, who was always quick to pass judgment, shouted: *"Ma Bill, è una apriscatole!"* ("But Bill, it's a can opener!").

In another example the kids were having an English lesson in which they were trying to learn the song "Twinkle, Twinkle, Little Star" in English. The English teacher, Joseph, first had the whole group of children sing the song in Italian and then went through it line by line with them in English. Next, he divided the kids into groups of four and asked them to sing the song in English, assigning a grade from 1 to 10 for their performance. I thought each group did pretty well, but Joseph was a tough grader and no group scored higher than 4 out of 10. Giovanna, who had been watching the lesson, suggested that I sing the song in English as a model. I had a feeling this was a setup, but I went ahead and, of course, Joseph gave me a perfect score.

"Now sing it in Italian," said Giovanna.

"Can the kids sing it once more in Italian for me?" I pleaded to Joseph.

They did so and I listened closely. Then I started up, but could not remember much after the first two stanzas and I stumbled over several words and then stopped singing altogether. Giovanna and the kids laughed loudly and Joseph called out my grade: "*Sotto zero!*" ("Below zero!").

By the end of the school year in early July, I had become very good friends with the kids, the teachers, and many of the parents. I was very pleased to be able to go along with the kids to elementary school in the fall. Of the original 21 children, 16 (5 children went to a different elementary school) were divided into four first-grade groups. I observed in a different group each day and often spent Fridays visiting the preschool teachers with their new group of three-year-olds. At first, in the elementary school, the children from our former preschool tried to claim me, by saying "Bill, belongs to us!" However, after a few weeks I got to know all of the other kids and by the time I left in December, the kids and teachers saw me as part of first grade! I remained a member of this group of children and their teachers all through elementary school.

However, one incident early in my time in first grade holds a special memory for me of my close friendships with the original preschool kids. It was mid-October 1996, and I had been with the first-grade kids for a little more than a month. I was in *Prima B* (first grade, group B). The teacher, Letizia, was moving some desks because children from *Prima A* were coming to visit the classroom. I was helping with this, when I felt the ground began to shake. It was an earthquake!

"We have to get the kids outside," said Letizia as she quickly left the room.

I assumed that I was to take care of the several kids in the room while she went to get those in the hallway, bathroom, or in *Prima A* where some were visiting. All of this happened in an instant and not only did the ground shake for several seconds, but it seemed to give way as if I were standing on Jell-o. I had been in a few "shake" earthquakes before, but this feeling of the ground giving away was new and frightening. I rounded up the five kids in the classroom and we went outside, where I saw groups of teachers and students gathered by the main gate. They were organized in classes and groups within classes. Some of the older children were frightened and crying, but the shaking had stopped by now. I looked at the taller buildings around the school but saw no damage.

As I got my kids with the rest of *Prima B*, I noticed several first-grade children go under a small enclosed area where bicycles were parked, to escape a steady drizzle. The teachers soon shooed them out—the point was to be away from anything that might fall down— and back to their group. Then one boy, Mario, from *Prima A* and also previously from the preschool where I worked, ran back toward the school. I started to go after him, but one of his teachers beat me to it and guided him back to his group.

"But I need my favorite pencil!" he protested.

"Are you crazy?" said the teacher. "We had an earthquake. You can get the pencil later."

By this time several kids who had been with me in the preschool and were in *Prima B* had pushed up close to me and grabbed my arms and legs as the teacher explained that we had just had an earthquake. After a few more minutes, things calmed down and the teachers let the children circulate among the first-grade group. Several kids from *Prima A, C,* and *D,* who had attended preschool with me, came running up and asked: "Bill, did you have an earthquake in your class too?"

2

"We're Friends, Right?"

· ·

Sharing and Social Participation in Kids' Culture

Richard and Barbara have been playing in the block area of the Berkeley preschool for several minutes. They are sitting near each other and building things with the small plastic blocks. They have not spoken and do not appear to be playing together. Up to this point their behavior would be seen as what many psychologists call parallel play.

Suddenly, Richard looks over at Barbara and says, "We're playing by ourselves."

"Just—ah—we're friends, right?" Barbara asks.

"Right," says Richard.

The two now coordinate their play and begin to build a house.

As this example shows, kids are social. They want to be involved, to participate, and to be part of the group. I saw little solitary play in my many years of observation in preschools. And when children did play alone or engaged in parallel play (a type, most common among toddlers, in which children play along side of but not really with each other in a coordinated fashion), it seldom lasted for long. They were soon doing things together.

I marveled at how kids worked together to get things going, like

Richard and Barbara did, and I shared in their joy when they marked their communal sharing with the oft-heard phrase, "We're friends, right?" Social participation and sharing are the heart of kids' peer culture.

But what exactly do I mean by kids' peer culture? I am using the term "peers" specifically to refer to that group of kids who spend time together on an everyday basis. My focus is on local peer cultures that are produced and shared primarily through face-to-face interaction. (Of course, local cultures are part of more general groups of kids, which can be defined in terms of age or geographical boundaries—for example, all three- to six-year-olds in the United States). Kids produce a series of local peer cultures that become part of, and contribute to, the wider cultures of other kids and adults within which they are embedded.

Much of the traditional work on peer culture has focused on adolescents and the effects (positive and negative) of experiences with peers on individual development. Most of this work has a functionalist view of culture; that is, culture is viewed as consisting of internalized shared values and norms that guide behavior.

In contrast, I take an interpretive view of culture as public, collective, and performative and define kids' peer culture as *a stable set of activities or routines, artifacts, values, and concerns that kids produce and share in interaction with each other.* As I noted in the preface, there are two basic themes in peer cultures: Kids want to gain control of their lives and they want to share that sense of control with each other. Throughout this book we will be considering many of the activities, routines, and artifacts of kids' culture and how children's participation in routines and use of artifacts reflect their shared values and concerns.

In this chapter I will concentrate on several activities or routines that are basic to sharing and control in kids' culture. Let's begin by returning to the play of Richard and Barbara.

"YOU'RE NOT OUR FRIEND":
THE PROTECTION OF INTERACTIVE SPACE

While Richard and Barbara coordinate their play and build a house together, another girl, Nancy (who entered the play area with Barbara), is sitting some distance away, watching them. Eventually she moves closer and sits next to Barbara, indicating her intent to play.

"You can't play," says Barbara.

"Yeah," agrees Richard.

Nancy gets up and moves farther away, where she sits down and watches again for a while. However, after a few minutes she gives up and goes to another area of the school.

I was uncomfortable when I saw kids reject the entry bids of their playmates. On some occasions such rejection seemed especially cruel.

Barbara and Betty leave the juice room together, move into the block area, and begin gathering blocks and toy animals. I realize this is a chance to observe the kids from the start of a play episode, so I quickly enter the area and sit on the floor near their play.

Barbara notices me and says, "Look, Bill. We're making a zoo."

"That's nice. You have lots of animals," I respond.

Betty looks up from her play and says, "Yeah, we're zookeepers. Right, Barbara?"

"Right," answers Barbara.

The two girls build small enclosures with the blocks, putting animals inside, and talking to each other about what they're doing. At one point, Betty sets some animals and blocks near me. "These are yours," she says.

Following the kids' lead, I build a small house and put some of my animals inside. I then notice Linda standing and watching us from the edge of the carpet that covers the block area. After a few minutes she enters, sits down next to Barbara, and picks up one of the animals.

Barbara takes the animal away from Linda and says, "You can't play."

"Yes, I can," Linda retorts. "I can have some animals too!"

"No, you can't," responds Barbara. "We don't like you today."

"You're not our friend," says Betty in support of Barbara's exclusion of Linda.

"I can play here, too," says Linda refusing to back down.

"No, her can't. . . . Her can't play. Right, Betty?" asks Barbara.

"Right," Betty confirms.

I'm bothered by this talk. However, in line with my ethnographic vow to not act like a typical adult, I try to stay out of the dispute. But then Linda turns to me and asks, "Can I have some animals, Bill?"

"You can have some of these," I say, offering her some of mine.

"She can't play, Bill," says Barbara, "'cause she's not our friend."

"Why not?" I ask. "You guys played with her yesterday."

"Well, we hate her today," snaps Betty.

I'm really set back by this retort and now even more uncomfortable. I'm somewhat relieved to hear Linda say, "Well, I'll tell teacher." She leaves and returns with a teacher who asks, "Girls, can Linda play with you?"

"No," Barbara replies. "She's not our friend."

"Why can't you *all* be friends?" asks the teacher.

Seemingly exasperated by this question, Barbara shrugs and mumbles, "No."

"Let's go outside, Barbara," suggests Betty. The two girls leave and go to the outside yard. Linda plays with animals near me for a while, but then gets up and goes into the juice room.

Most adults would be troubled by the behavior of Barbara and Betty. Why wouldn't they let Linda play? Why couldn't they, as the teacher suggested, "all be friends"? To answer these questions we need to suspend our adult perspective of sharing and friendship and consider things from the kids' point of view.

Let's go back to the first example of Richard and Barbara, the house builders. Initially, their play was parallel, uncoordinated, and lacking in a shared focus. But once they began to play together, they quickly agreed that they were friends. For young children, the kids you are playing with are your friends, while those not playing are often seen as a threat to friendship. But why is this the case? Why are the children so protective of their shared play?

As adults we can easily suspend our interactions and conversations to handle brief disruptions like phone calls or a crying child and pick up where we left off. It's not so easy for three- to five-year-olds. Establishing and maintaining peer interaction are challenging tasks for kids who are in the process of developing the linguistic and cognitive skills necessary for communication and social interaction. Furthermore, the social ecology of most preschools increases the fragility of peer interaction. A preschool play area is a multiparty setting much like a cocktail party with lots of clusters of kids playing together. Kids know from experience that at any moment a dispute might arise over the nature of play ("Who should be the mother and who the baby?" "Should the block go this way or that?"), other kids might want to play or take needed materials, or a teacher might announce "clean-up time." Kids work hard to get things going and then, just like that, someone always messes things up.

The children's desire to protect interactive space is not selfish. In fact, they are not refusing to share, rather *they want to keep sharing what they are already sharing.* Consider again the example of Betty, Barbara, and Linda. Betty and Barbara entered the block area together and quickly established a play theme of building a zoo and being zookeepers. I sat nearby and made no attempt to enter, intervene, or question them about their play. Given my established status as an adult friend who did not intervene in or try to direct their play, the girls told me what they were doing and offered me play materials. Linda, on the other hand, was seen as threatening. She entered without invitation and her bid to play was quickly resisted.

Linda's insistence that she had a right to play only increased the other girls' perception of her as threatening to their play. Neither my attempt to aid Linda's entry by reminding Betty and Barbara that they had played with her previously, nor the teacher's suggestion that they "all be friends" was successful. In fact, Betty and Barbara were annoyed by the teacher's intervention and abandoned their play. Finding herself alone with me, Linda quickly left to seek out a more desirable playmate.

My point is not that adults should always refrain from intervening in children's peer disputes. Surely, adults (especially preschool teachers) need to protect children from physical and emotional abuse by their peers. Even in this mild case of rejection, the teacher responded appropriately to Linda's request for help. Also, it was good for Betty and Barbara to be reminded that the needs of others to enter their play differed from their own desire to keep the play intact. Finally, the behavior of Betty and Barbara if frequent in its occurrence and ignored by teachers could lead children to become disrespectful of their peers. Yet we adults should not expect that children can easily appreciate such advice, nor should we assume that we can easily teach kids how to be social. Adults can be helpful, but children often *collectively teach each other* how to get along.

Here's another example. The play again involves Betty, who this time is at the outside sandbox with Jenny. The girls pretend to prepare dinner and put sand in pots, cupcake pans, and teapots as I sit nearby, watching. Suddenly Debbie approaches and the following occurs:

Debbie comes up to the sandbox and stands near me, closely watching the other two girls. After watching for about five minutes, she circles the sandbox three times and stops again and stands next to me. After a few more minutes of watching, Debbie moves to the sandbox and reaches for a teapot. Jenny takes the teapot away from Debbie and mumbles "No." Debbie backs away and again stands near me, observing the activity of Jenny and Betty. Then she walks over next to Betty, who is filling the cupcake pan with sand.

Debbie watches Betty for just a few seconds, then says: "We're friends. Right, Betty?"

Betty, not looking up at Debbie and continuing to place sand in the pan, says, "Right."

Debbie now moves alongside Betty, takes a pot and a spoon, begins putting sand in the pot, and says, "I'm making coffee."

"I'm making cupcakes," Betty replies.

Betty now turns to Jenny and says, "We're mothers. Right, Jenny?"

"Right," says Jenny.

The three "mothers" continue to play together for about 20 more minutes, until the teachers announce clean-up time.

In this example we see how Debbie overcomes the resistance of the other kids and successfully enters their play. She does this by employing what I call *access strategies*—procedures for gaining entry into ongoing interaction. First, Debbie merely places herself in the area of play, a strategy I call *nonverbal entry*. Receiving no response, Debbie keeps watching the play, but now physically circles the sandbox (what I term *encirclement*). Some child researchers refer to Debbie's actions as "onlooker behavior" and argue that it is an indicator of timidity. However, it is important to observe access attempts within their social contexts and not rely on short, arbitrary time samples when studying children's play, which has been the case in much research on young children's play. Although onlooker behavior may occur, it can often be part of more complex sequences of behavior. Observing entire episodes of interaction, I found that access attempts often involve a series of strategies that build on one another.

In this case, Debbie, when stationary and on the move, carefully makes note of what the other kids are doing. With this information she is able to enter the area and do something in line with the other kids' play (that is, pick up a teapot). Although often a successful access strategy, it is initially resisted in this instance. Not giving up, however, Debbie watches some more, again enters the area, and makes a verbal reference to affiliation ("We're friends, right?"). Betty responds positively, but does not explicitly invite Debbie to play. Debbie then repeats her earlier strategy of doing something similar to what the other kids are doing, this time verbally describing her play ("I'm making coffee"). Betty now responds in a way that includes Debbie in the play, noting that she is also making something (cupcakes). She then goes on to define further the new situation by saying "We're mothers," which is confirmed by Jenny. Debbie is now clearly part of the play.

Although Debbie is eventually successful, one might wonder why she simply did not go up and say "Hi," "What ya doing?," or "Can I play?" I have found that preschool children rarely use such direct strategies. One reason is that they call for an immediate response, and that response is very often negative. Remember my earlier point about the protection of interactive space. Kids fear that others may disrupt the cherished but fragile sharing they have achieved. Direct entry bids like "What ya doing?" or "Can I play?" or the frequently heard "You have to share!" actually signal that one does not know what is going on and, therefore, might cause trouble.

Again, we see why it is important for adults to take kids' perspectives. What might seem like selfish behavior is really an attempt to keep sharing. Further, by actively confronting resistance to their access attempts, children acquire complex strategies that allow them to enter and share in play. There is one more point. The access skills the kids develop in this multiparty setting are clear precursors to adult skills that are used in similar situations. Picture yourself at a party. Let's say you have just arrived, gone off to get a drink, been to the bathroom, or some such thing. Now, like the kids in the preschool, you don't want to remain alone.

What do you do? Go up to a group and say "Hi," "What ya talking about?," "Can I talk too?" Probably not. You're more likely to stand near a group, listen, figure out what they are talking about, and make a relevant contribution to the conversation. In short, you do pretty much what Debbie did in the previous example. There is one difference, however. We adults are not likely to tell the guy who bursts in on a conversation that he "is not our friend" or to "beat it." We might want to, but we send more subtle signals—like ignoring what he has to say. As grownups we have learned tact (though it does not always work as well as we might like).

The examples of protection of interactive space I have presented so far are all from the Berkeley preschool. Over time in this school, children became more adept at gaining entry and there was less need for them to protect their play. I found a similar temporal pattern in the protection of interactive space in all the American schools I studied. However, things were somewhat different in Italy.

In Modena, where the kids had been together for two and a half years before I arrived, protection of interactive space was rare. It occurred only in play where there was not enough space for additional kids to join in or when a child entered ongoing play in a disruptive way. The latter was rare and usually involved boys disrupting girls' play just for the fun of it.

In Bologna, in the first several months of the school term, the kids frequently protected their interactive space while playing in the large inside playroom in the school. Unlike the American schools I studied, in which the playrooms were all divided into small subareas by partitions (such as bookshelves or cupboards), the main playroom in the Bologna preschool had one large area of open space surrounded by chairs placed against the walls of the room. As a result, the kids usually carried play materials to various places in the room and formed small groups that were vulnerable to the attempts of other kids to gain access, as well as to other potential disruptions.

In an example from my field notes, Bruna, age three, and Cinzia, age four, are building a house with Legos near the center of the room. The girls have placed a number of toy animals inside the house but become frustrated by several disruptions, including entry attempts by three other kids. So the girls move over by the chairs along one of the walls of the room. The chairs are box shaped, with equal space above and below the seat. When they reach the chairs the girls put their toy house and the animals under a chair and sit in front of it, hiding their play from the direct view of others.

Bruna says, "We're playing here."

"Nobody comes here," adds Cinzia.

A few minutes later, Gina approaches and sits in the chair next to the one under which the other girls are playing. At first, Bruna and Cinzia ignore Gina, but when she attempts to sit on the floor and reach for an animal, Cinzia pushes her and says, "Go away."

Gina insists that she can play, but Bruna and Cinzia say she cannot. Bruna moves to block off Gina and she and Cinzia continue to play. Gina does not give up, however, and continues to reach for animals and says she has a right to play. Finally, Bruna and Cinzia abandon their toy house and animals and move to another area of the room. Gina plays with the toys briefly, but then also goes off to find other playmates.

While this example is similar to many instances I observed in the United States, I later observed several episodes of peer play in Bologna that were quite different. In these episodes, which normally occurred in the outside yard, kids did not hide their play from others. In fact, they frequently announced what they were doing and allowed other kids to participate. Consider the following videotaped example.

Carla, about five years old, and Federica, about six, are sitting on the steps in front of a small bathhouse in the outside yard of the preschool. The bathhouse is no longer in use because the wading pool near the school is closed. Carla picks up a rock and begins to rub it against the steps. Federica finds a rock and joins Carla in the activity.

Carla then decides that she wants to move a much bigger rock and set it on the steps. She gets me to help and once the rock is in place, the two girls rub their smaller rocks against it, making a white powder.

"We're making it all white," says Carla, referring to the powder.

The girls rub harder now, making more powder. Carla places her hands in the powder, holds them up, and says, "They're all white!"

Federica now gets her hands all white, and both girls laugh, pleased with their creation.

The two girls continue their play for 10 minutes and then four other girls (Flora, Bianca, Giovanna, and Viola, all about five years old) come over and sit on the steps near them. The four girls pretend they are riding a bus and make "motor" sounds. Carla and Federica pay little attention to them while they continue making their powder. After a few minutes Carla turns to the other girls, holds up her hand and says, "Look, I have my hand all white."

"I also want my hand all white," says Bianca.

"Come here with me," directs Carla.

Bianca comes over to Carla and holds out her hand. Carla takes it and places it on the stone, covering it with powder.

Now the other girls crowd around, trying to get a turn.

"Me too!" shouts Giovanna.

"Me too!" echoes Flora.

"Wait," commands Carla. "One at a time!"

Carla finishes Bianca's hand and Viola pushes forward, "Me too!"

"First her," says Carla, pointing to Giovanna.

"I'll do it with you," Federica tells Viola. She places Viola's hand on the rock, but does not get much powder on it. Carla then takes Viola's hand and rubs it harder so that it is covered with powder.

Giovanna now pushes Viola away, "Also me."

Carla does Giovanna's hand and then Flora's. After she has her hand done, Flora goes back and reaches for more.

"Enough!" Carla says visibly upset. "You're taking it all!"

Bianca now returns, pushes Viola aside, holds up both her hands, and says, "I want it like this on both hands."

"Me too!" shouts Viola.

"Enough!" replies Carla. "Go away."

The two girls return and join Giovanna and are not allowed to touch the powder. They again pretend to drive a bus, while Carla and Federica continue to make their powder. After about 15 minutes it is time to go inside.

In this example we see a basic difference from the American examples of protection of interactive space. The Italian girls actually described their play to their peers and allowed them to participate. However, the newcomers' participation was severely restricted, and when they challenged the kids in charge of the play they were excluded. So we see that the Italian children shared the American children's concern with maintaining shared activities. However, the Italian kids displayed more confidence in their communicative abilities to maintain control of their play without closing themselves off from the attention of peers.

"WE'RE BIGGER THAN ANYBODY ELSE"

As I lean against the back wall of the outside yard in the Berkeley preschool observing kids play on the nearby climbing bars, I think about how often the kids play on the bars when outside, and how, when climbing high on the bars, they can see over the walls and beyond the confines of the school. This reflection gets me thinking about how this reversal of physical perspective—the children looking down on rather than up to adults—empowers kids.

Then I hear Laura, who has climbed nearly to the top of the bars with Christopher, yell down to Vickie, who is standing with Daniel near the base of the bars. When Vickie looks up, Laura shouts, "We're bigger than you!"

"Oh, no you're not," retorts Vickie, as she begins to climb to the top.

Daniel follows close behind and both call out, "No, you're not!"

As Vickie and Daniel get near them, Laura and Christopher move to the very top level and Christopher says, "We're higher now. Right Laura?"

"Right," Laura responds. "We're higher than anybody else!"

Vickie and Daniel now climb to the highest level, and Vickie shouts, "We are higher now too!"

Laura then repeats in a measured cadence, "We are higher than anybody else!"

Now all four kids chant in unison, "We are higher than anybody else! We are higher than anybody else!"

After several repetitions, the kids slightly alter the chant and yell, "We are bigger than anybody else! We are bigger than anybody else!"

Several other kids hear the chant and head for the bars to climb up and join in. I look up and realize that the kids are at this moment taller than I am. Unlike their playmates who are now scrambling up the bars, I cannot so easily take up the challenge. I'm constrained by my adult body. I'm too big to make myself bigger.

Being bigger is valued in the peer culture and kids collectively share and display this value in routines like the one described. In all the schools I observed, I found that the children prefer to play in areas where indeed they are bigger and looking down at others, especially adults. Climbing bars and other structures is also fun because they are designed for children and challenge their physical skills. So kids often embellish their play in these areas by doing tricks on the bars, going down the slide backward, and so on. In doing such tricks the kids often call out for the attention of peers and adults.

In this vein the kids also gain some autonomy by reaching outside the boundaries of the school and the direct control of the teachers. In Bologna, in Modena, and in the Indianapolis and Bloomington Head Start programs, the kids relished calling out to adults walking by their school, frequently engaging them in conversation. At times they asked the adults to watch them climb high on the bars, go down the slide, swing fast and high, or simply run and jump.

In Berkeley, the kids in the morning session engaged in a routine that contained all these elements of size, autonomy, and reaching out to influence adults beyond the school boundaries. The following is drawn from my field notes when I first observed the routine.

It was a beautiful November morning. In fact, it was much too nice a day to remain inside the school. So, like most of the kids, I decided to spend time in the outside yard. Once outside, I noticed Michelle, Jimmy, and Dwight moving toward the sandpile, and I quickly joined them. While the bright sun warmed the back of my neck, I sat in the sand watching the kids digging.

Suddenly, I heard a loud shout and turned to see Denny, Leah, and Martin on the climbing bars. Denny was shouting and pointing over the back fence of the yard toward Kelly Street. Something must be going on out there that the kids could see from high in the bars. Then Leah shouted: "It's him! It's him!" My curiosity was aroused. As I got up to go look, Michelle, Jimmy, and Dwight abandoned their shovels and ran past me to the bars. Just as I reached the bars, several kids began shouting: "Garbage man!" "Garbage man!"

I moved beyond the bars, peered out over the fence, which was about neck high for me, and did indeed see a garbage man. In fact, there were two garbage men out there, along with a large garbage truck. One of the men sat behind the wheel of the truck which he had—I assumed—backed up in front of the dumpster near the apartment building across the street from the school. The other man had moved to the rear of the truck and seemed to be attaching the dumpster to a lift. He then yelled, "Ready," to his partner, and the dumpster began to rise from the ground accompanied by a loud whirring.

The kids were very excited and were imitating the noise of the truck lift: "Whirr!" "Whirr!" "Whirr!" I was surprised to see that there were now more kids on the bars: ten in all, with one more, Barbara, climbing up. I looked around the yard and noted that all but two of the children who were outside were now on the bars. As the dumpster reached its apex and the trash tumbled into the truck, the kids seemed to reach their own peak of excitement. They waved and "whirred" in near perfect unison. At exactly this point, the garbage man outside the truck looked up and waved back to his admirers. The lift then lowered quickly and the dumpster hit the ground with a loud bang. The outside man unhooked the dumpster and joined his partner in the truck. The kids continued waving and shouting "Garbage man!" as the driver pulled the truck away, gave a beep of the horn, and steered the truck down the street to the next stop, far beyond the sight of the kids.

With the garbage truck out of sight, I noticed that most of the kids had left the bars and returned to other play activities. I was surprised that I had never noticed the garbage man routine before. I wondered how often it occurred. Did the teachers know about it? Will it recur tomorrow?

The garbage man routine did indeed recur the next day, and eight kids participated. The following day it occurred again with five kids involved. In all, the routine was enacted every day for the 110 days that I checked for its occurrence over the remainder of the school term. The number of kids participating in the routine ranged from two to thirteen, and all but two girls participated in the routine a least once. Two of the three teaching assistants and the head teacher were aware of the routine when I asked them about it. Only the teacher had paid much attention, and she remarked about how nice it was that the garbage man always waved to the children while making his pickup.

The garbage man routine shares several characteristics with other elements of peer culture we have discussed or will discuss. First, there is the sharing of excitement and joy that we saw in the kids' chants about being bigger. Second, the routine involves a group production that builds and reaches a climax at a predictable moment. This sequential pattern will also be apparent in approach-avoidance play and spontaneous fantasy, which we discuss later. Finally, the routine emerges in reaction to something that has special appeal to the kids: the loud "whirring" and "clanging" of a big and, to the children's eyes, beautiful machine.

At a more abstract level, "garbage man" is different from other elements of peer culture. For in this activity the children literally *reach out beyond* the physical boundaries of the preschool to the adult world, and transform a mundane event (the collection of garbage) into a routine of peer culture that they collectively produce and enjoy. And, at an even deeper level, the routine is significant because the kids are *successful at procuring the participation of adults in an event that the children create and control and of whose significance adults have only a surface recognition.*

Although my research in the Berkeley preschool was limited to one year because I took a job in another part of the country, I did return during the next year for some limited observations. The morning group from the year before now attended afternoon sessions, and there was a new morning group. Because the garbage was collected in the morning on Kelly Street, I was eager to see if the new kids had noticed the garbage collectors and had developed a routine similar to that of their counterparts from the year before. On my first day back I anxiously waited at the back fence for the garbage truck to appear. I was a stranger to the three boys who were playing in the bars, and they paid me scant attention as I awaited the truck. Finally, the truck arrived and backed up to the dumpster. Then I heard it, shouts of "Garbage man!" from the kids in the bars. Two boys and two girls quickly joined their peers on the bars and the routine was enacted: the whirring, the shouts, the wave of the hand, and the beep of the horn. Although I cannot be sure, I suspect "Garbage man" might still be alive and well at the Berkeley preschool.

"WATCH OUT FOR THE MONSTER":
APPROACH-AVOIDANCE PLAY

One morning in the Berkeley preschool, four boys (Denny, Jack, Joseph, and Martin) were playing in the upstairs playhouse. At one point, the boys started wrestling and giggling on the bed. As they untangled, Joseph pointed at Martin and yelled: "Watch out for the monster!"

"Yeah, watch out!" yelled Denny and Jack as they and Joseph ran downstairs as if fleeing in fear from Martin.

Martin was bewildered by this turn of events. He walked over to the stairway to see where his friends had gone and, then not seeing them, returned near the bed and peered down into the school.

Meanwhile, the other three boys huddled together in the downstairs playhouse against the wall near the stairway, out of Martin's view. They laughed and Denny pushed Jack toward the stairway, "Go see where the monster is."

Jack crept cautiously out of the downstairs playhouse, looked up, saw Martin looking down, and ran back inside screeching, "Here he comes!"

Martin, still confused about what was happening, moved slowly down the stairs. He eventually reached the bottom, turned the corner, and saw the other boys. The three boys then screamed and ran back upstairs. As they passed Martin, they bumped into him, spinning him around. Looking back at Martin at the top of the stairs, the boys yelled, "You can't get us, monster!"

Martin now began walking mechanically like a robot and pursued the other boys back upstairs. When he got to the top of the stairs, the other boys again ran by him, fleeing in mock fear. This cycle was repeated several additional times before the play ended with clean-up time.

After viewing and transcribing this play episode, which I had videotaped, I recorded several things in my theoretical notes. First, I was taken by how Martin was thrust into the role of monster by the other boys and how it took him a while to realize what was going on. Second, it was clear that the other boys were only pretending to be afraid of Martin, but nevertheless the play generated a good bit of excitement and tension. Finally, once Martin realized he was identified as a monster, he embraced the role and several cycles of fleeing and chasing ensued.

Shortly after the episode occurred, I was sitting in the outside sandpile of the Berkeley preschool with Glen, Leah, Denny, and Martin. Rita, who was wearing a dress with an apple print, walked by us. Glen yelled, "Hey, there's the apple girl!"

"Watch out! She'll get us!" shouted Denny and he and the others ran toward the climbing bars.

Rita then spun around, raised her arms, shaping her hands into claws, and ran after the other kids in a menacing fashion. When Rita got near her intended victims, they ran around her and back to the sandpile.

Rita did not pursue them into the sandpile, but rather circled around it. As she passed by the second time, the other kids again ran up behind and past her toward the bars, yelling, "You can't get us Apple Girl!" Rita again pursued them to the sandpile, and this routine was recycled several more times.

After recording this event in field notes, I wrote in my theoretical notes that it was much like the earlier routine. Here a monster (or threatening agent) was created or identified and approached and avoided. There was an addition in that the sandpile was treated as a home base for the threatened children. At this point I began to refer to this play routine as approach-avoidance.

Let's consider one more example of approach-avoidance displayed by the Berkeley kids before examining the structure and significance of the routine in more detail. Three children in the afternoon group, Beth, Brian, and Mark, are playing on a rocking boat in the outside yard. After about 10 minutes of rocking, Beth notices Steven, who is walking at some distance from the boat with a large trash can over his head.

"Hey, a walking bucket! See the walking bucket!" shouts Beth.

Brian and Mark are facing the opposite direction and do not see Steven. "What?" says Brian.

"A walking bucket. Look!" says Beth as she points toward Steven. Brian and Mark now turn and see Steven.

"Yeah," says Brian. "Let's get off."

The three kids stop the boat, jump down to the ground, and with Mark leading the way, move slowly toward Steven.

Steven can't see the other kids coming as he stops walking and stands in an area where large wooden blocks are stored. When they reach Steven, Mark and Brian push the trash can and start to raise it above Steven's head.

"You," shouts Steven and he flips the trash can off his head.

"Whoa!" yells Brian and he, Mark, and Beth run back toward the rocking boat.

Steven starts to put the trash can back on his head, but when he sees the other kids running he drops it to the ground. He then runs toward the rocking boat, flailing his arms in a threatening manner.

Brian, Mark, and Beth pretend to be afraid, screech, and move to the far side of the boat. Steven stops at the vacant side of the boat and rocks it by pushing down on the boat with his hands. However, he does not climb onto the boat, nor does he try to grab the other kids.

Steven then returns to the dropped trash can and puts it back over his head. Brian, Mark, and Beth watch from the boat, giggling and laughing. Once Steven has the trash can back over his head, Mark says, "Let's kick him."

Mark and Brian jump down from the boat and move toward Steven who still has the trash can over his head. However, it appears that Steven expects the other kids to return so he stays near the block area. Beth remains behind on the boat.

Mark reaches Steven first and kicks at his legs but misses them and instead kicks the bottom of the trash can. Brian now runs up and also kicks at Steven but clearly misses. The two boys then run back to the boat just as Steven raises the trash can off his head. Steven flips the trash can to the ground just as Brian and Mark get back on the boat with Beth. Steven takes on a threatening stance but remains silent and does not move toward the boat. Instead he places the trash can back over his head and walks around the yard, moving farther away from the other kids to the end of the sandpile opposite the boat.

The other kids now begin to rock the boat very fast, "Whee! Faster! Faster!" shouts Beth.

Steven is still some distance from the boat, but now he begins to move in the direction of the other kids. It is not clear how Steven knows where he is going, because he can use only what he sees on the ground directly in front of his feet to guide him.

"Hey, he's coming!" yells Beth.

"Hey, you big poop butt!" taunts Brian.

All the kids laugh, and Beth yells, "Hey you big fat poop butt!"

Steven ignores these taunts and continues to walk around the yard. Beth now jumps from the boat and runs toward Steven with Brian close behind. Mark also has left the boat but is trailing the other two. As Beth nears Steven, she veers off to the left, while Brian runs up to Steven and pushes the trash can. Mark arrives just as Steven flips off the can and shouts, "I'll get you!"

Steven chases Mark and Brian back toward the boat but takes a circuitous route, which allows the boys to easily make it to the home base. Steven again pushes the vacant side of the boat and then returns to the trash can. Brian and Mark rock on the boat, watching Steven place the trash can back over his head. Beth has now left the game and is playing elsewhere.

Once Steven has the trash can over his head and is again walking around, Brian jumps from the boat and runs right to Steven. Just as

Brian arrives, Steven flips off the trash can and grabs him. Mark, who was following Brian, now returns to the boat and watches as Steven and Brian get into a mild tussle. This physical conflict, which is rare in approach-avoidance play, leads to the intervention of a teacher and the end of the routine.

This videotaped example of the approach-avoidance routine nicely displays its basic features. The routine always contains a threatening agent (such as a monster, wild animal, or, in this case, a "walking bucket"), who is both approached and avoided. The routine has three distinct phases: identification, approach, and avoidance.

In the identification phase the children create or discover and mutually signal a threat or danger. This phase is important because it serves as an interpretive frame for the activities that follow. That is, identification of a shared threat signals that the approach-avoidance routine is under way and that emerging activities should be interpreted in line with the play theme.

In some cases, one or more children adopt the role of threatening agent. However, the threatened children must accept or ratify children who embrace the threatening agent role for identification to occur and for the routine to continue. Sometimes children who pretend to be monsters, evil villains, mad scientists, or other types of threatening agents are ignored or rejected (for example, "Go away. You're scaring us"). In these cases, there is no identification and the approach-avoidance routine fails to materialize. In many cases, children are literally thrust into the role of threatening agents, as we saw in all the examples we have considered to this point.

In the approach phase, the threatened children advance cautiously toward the source of the danger. During this approach, the threatened agent is sometimes disabled in some way, as we saw in the walking bucket, where Steven could not see the other children approaching him. However, more often threatening agents pretend not to see or hear those approaching until they are very close, almost to the point that the threatened children nearly or actually touch them. Such inattention heightens the emotions of the children and the tension in the routine.

In the avoidance phase the threatening agent enables himself or herself, often with an evil growl or scream and threatening gestures, the threatening children flee with ample display of feigned fear, and the threatening agent chases after them. In most cases the threatened children escape to certain areas that serve as a "home base" (for example, the sandpile or the rocking boat).

Eventually the attacker moves away and the danger diminishes. At this point the routine might end, but most often the threatened children initiate a new approach phase. In some cases the approach and avoidance phases are repeated several times, with more participants entering and exiting the routine.

I have not found in my research that the same child was adopted or was repeatedly thrust into the role of threatening agent. The walking bucket episode was the only time Steven ever put a trash can over his head or took on the role of threatening agent. The children prefer to be threatened because in this role they are the ones who control the initiation of the play in the identification phase, who are ultimately protected by reaching home base, and who frequently embellish the routine (for example, by exaggerating their fear with loud shrieks and screams and by taunting the threatening agent with insults).

Although I identified the approach-avoidance play in Berkeley, I discovered and recorded the routine being enacted in spontaneous or formal modes in all the preschools I studied. For example, in the Indianapolis Head Start, the kids frequently played a run and chase game they called "Freddy." Freddy is an evil character from the *Nightmare on Elm Street* horror movies. In the play a child often volunteered to be Freddy and other children approached and then avoided this evil villain who pretended to have long, razor-sharp fingernails.

I was surprised to learn that such young children watched these horror films and commented to Zena (who was pretending to be Freddy) and one of her friends, Ramone (who had been fleeing from Zena), that they shouldn't watch such movies because they're "too scary." Both children scoffed at my concern and Ramone pointed out that Freddy was not real, but just a man in a costume. I persisted, noting that such movies gave me bad dreams. Zena responded that she didn't have bad dreams about Freddy, but rather of a dog that chased and tried to bite her. Zena and Ramone left little doubt about their ability to distinguish make-believe and real threats, and they displayed mature and sophisticated reasoning for children of their age, which in some ways reflected the reality of the challenging economic circumstances in which they lived.

LA STREGA AND GAINGEEN: APPROACH-AVOIDANCE PLAY IN OTHER CULTURES

The Italian preschoolers I studied, like their American counterparts, engaged in spontaneous approach-avoidance routines. However, the Italian kids also produced a formal game they referred to as "*la Strega*" ("The Witch") that has a participant structure very similar to the approach-avoidance routine. The Italian kids' fascination with witches is no doubt related to the mythical character "la Befana." La Befana, who is believed to have originated in Southern Italy, is a witch who flies on a broom and brings presents to children on January 6, at Epiphany. (Epiphany is a festival of the Catholic Church commemorating the coming of the Magi as the first manifestation of Christ to the Gentiles.) According to the legend, the three wise men stopped to ask la Befana for directions on their way to Bethlehem. They also invited her to join them. However, la Befana said she was too busy sweeping and sent them away. She was soon filled with remorse and set out to follow the Magi. She couldn't find them and has been flying around Italy ever since, looking for the Christ child. She leaves presents at the house of every child in case one of them is the Savior. The legend has been altered in modern times, with parents warning their children that la Befana does not leave presents for bad children. She is said to slide down chimneys on her broom, leaving presents and candy in the shoes of good children, while bad children get switches and lumps of coal.

The first step in playing *la Strega* is getting a playmate to agree to take on the role of the witch.

Cristina, Luisa, and Rosa (all about four years old) are playing in the outside yard of the Bologna preschool. Rosa points to Cristina and says, "She's the witch!"

Cristina does not answer but seems reluctant.

"Will you be the witch?" asks Luisa.

"OK," Cristina agrees.

Cristina now moves away from the other two girls and places her hands over her eyes. Luisa and Rosa slowly move closer and closer to Cristina, almost touching her. As they approach, Cristina repeats, "*Colore! Colore! Colore!*" ("Color! Color! Color!").

Luisa and Rosa draw closer with each repetition. Sensing they are very near her, Cristina shouts "*Viola!*" ("Violet").

Luisa and Rosa run off screeching, and Cristina, with her arms outstretched in a threatening manner, chases after them. Luisa and Rosa now run in different directions, and Cristina pursues Rosa. Just as Cristina ("*la Strega*") is about to catch her, Rosa touches a violet-colored object (a toy on the ground that serves as home base).

Cristina now turns to look for Luisa and sees that she has also found a violet object (the dress of another girl). Cristina again closes her eyes and repeats: "*Colore! Colore! Colore!*" The other girls begin a second approach and the routine is repeated, this time with "gray" as the announced color. Rosa and Luisa again find and touch correctly colored objects just before Cristina is about to catch them.

Cristina then suggests that Rosa be the witch and she agrees. The routine is repeated three more times with the colors yellow, green, and blue. Each time the witch chases but does not capture the fleeing children.

We see that the Italian children have formalized the main characteristics of the approach-avoidance routine (a threatening agent who is approached when disabled and avoided when empowered, feigned fear on the part of threatened children, the security of a home base, the buildup and release of tension, and possibilities for repetition and embellishment) into a game that they can enact at any time. In fact, in my second year at the Bologna preschool I learned that the kids had created an interesting variation of *la Strega*.

I wanted to videotape an instance of the game and I asked the kids to play "*la Strega*" for me. A girl, Martina, asked, "Do you want '*la Strega colore comando*' or '*la Strega bibita*'?" I immediately realized that the version of "*la Strega*" I had seen and recorded in field notes the year before was "*colore comando*," but I was curious to learn more about "*la Strega bibita*." I knew that "*bibita*" was the word for soft drink or refreshment. However, I was not sure how a soft drink would be part of the approach-avoidance structure. So I said, "Show me '*la Strega bibita*.'"

The kids agreed and Martina volunteered to be the witch. Maria huddled the kids into a group and each child whispered into her ear. I moved close and could hear that they were telling her different flavors of soft drinks (orange, cherry, lemon, and so on). When a kid whispered the choice of another, Maria told her or him to select a different flavor. Finally, all the kids had a different soft drink. They then knelt on the ground in a line, facing away from Martina. Martina approached and walked up and down the line several times. Finally she stopped behind a girl, Elena, and tapped her on the back.

"Who is it?" asked Elena.

"La Strega," answered Martina.

"What do you want?" said Elena.

"Una bibita," Martina responded.

Then Martina backed away as all the kids got to their feet and approached her moving abreast of one another. After the kids took several steps, Martina commanded them to keep their line straight and they did as they were told. Finally, when they were a few feet from Martina, she shouted out a flavor, "Orange!" Rita had selected "orange," and she started to run, but stumbled and Martina grabbed her quickly. The rest of the kids were very unhappy with this turn of events and criticized Rita for her clumsiness. Martina said that they should back up a bit and approach again and she would select a different flavor. The other kids did so and this time she shouted "Grape!" and Luca took off, running quickly with Martina in hot pursuit. She chased Luca around the yard with the other kids in the group chanting "Luca! Luca! Luca!" Luca could run faster than Martina and circled the yard and got back to the group before the witch could catch him. He was welcomed with many cheers and pats on the back by the other kids. Martina was a bit unhappy and complained that Luca had cheated in taking the particular route he did. The other kids dismissed this complaint and Martina gracefully accepted that Luca had escaped her and suggested that a new witch be selected to continue the play.

The Italian children's approach-avoidance play is impressive in that they have formalized the routine into a game with general rules (*"la Strega"*) and over the course of a year invented a variation on the original game. The new game (*"la Strega Bibita"*) had a number of interesting features that illustrate the innovative nature of children's peer culture. First, instead of the children identifying and approaching the witch, the witch first approaches a group of children and singles out a particular child to initiate the routine. This initial phase involves collaboration among the group of threatened children to select a flavor of soft drink (something children like) and a buildup of tension as the witch approaches the group and walks up and down before selecting a child to ask for a soft drink. The child playing the witch has a great deal of freedom to decide when to actually initiate the play as she passes all the children several times before making a selection. Once the witch asks for a soft drink, the other children approach slowly in a straight line, getting very close to the threatening agent. The witch then shouts out her selection of a flavor of soft drink and a particular child is *thrust from the group* and attempts to escape the witch. Although alone, this child is supported and cheered on by playmates to escape the witch. In this variation, then, the home base becomes the group of threatened children themselves and the individual wins out over the witch by re-uniting with the group.

Other variants of approach-avoidance routine have been reported in cross-cultural studies of children's play. One example can be seen in the work of the anthropologist Kathleen Barlow, who studied the Murik, a fishing and trading society of Papua New Guinea. Barlow found that the Murik believe in a number of spirits who display hu-man-like tendencies "toward mischief, deceit, and irritability." One of these spirits is *Gaingeen*, who appears sporadically in the village to chase and threaten children. Gaingeen never speaks, but rather con-veys his intention through threatening gestures and shaking the spears and sticks he always carries with him, as we can see in Figure 3.

Parents often use the threat of Gaingeen to get young children to stop crying or to dissuade them from undesirable behavior (for ex-ample, a grandmother recalls wandering toddlers from doorways with the threat "Eeee! Gaingeen! Gaingeen!"). As children grow older, they learn that Gaingeen does not come every time he is called. Nonethe-less, he might come, and the young children always look around to see if he will appear when caretakers call his name.

Eventually, when children are seven years old or so, the secret of the masked figure is revealed and the children discover that Gaingeen is an adolescent boy wearing a costume. Despite this demystification, Gaingeen remains an important figure in play and learning throughout childhood.

In her analysis Barlow describes a strikingly similar type of ap-

FIGURE 3 Gaingeen arrives in the village.

proach-avoidance play among children of the Murik in response to
Gaingeen. For example, preadolescents look forward to Gaingeen's
appearance in the village and run up close to him, often taunting the
bogey with insults: "Gaingeen, you dirty dog, you!" But they quickly
run for cover when Gaingeen charges after them. Such routines are
repeated many times during Gaingeen's visits to the village.

Younger kids keep their distance and watch but do not engage in
the approach-avoidance routine. Later, however, these kids re-create

FIGURE 4 Murik children making gaingeen costumes.

in their play the events they observe. In this play, seven- and eight-year-olds make Gaingeen costumes for younger children (four- to six-year-olds). Once the little ones are walking around pretending to be Gaingeen, the older children playfully approach and tease them and then run away in feigned fear when the younger ones threaten them. Figure 4 shows older children making Gaingeen costumes for the younger ones.

SHARING AND CONTROL

In all the play routines of peer culture discussed in this chapter, we can see the general pattern of sharing and control. In protecting interactive space, the kids establish shared play and then work hard to keep control of the often fragile interaction. Thus, their resistance to other kids' entry bids is not a refusal to share but an attempt to keep control of their play, *to keep sharing what they are already sharing*. In the process, kids sharpen their developing interactive skills as they build complex play activities and acquire needed access strategies to show that they can fit into the play. And all the kids end up having fun!

Climbing high on bars, playhouses, and other structures is also fun and it gives kids a sense of control over adults. For during these moments looking down on adults, kids are really bigger, and they display this sense of control with chants and taunts. The kids use their bodies to make the most of spaces that cannot be easily shared by adults.

We saw a wonderful extension of "being bigger" play in the garbage man routine created by the Berkeley kids. There they used their size and ability to climb high on the bars to reach out beyond the confines of the preschool to create a special routine in which adults, going about their everyday activities, are incorporated into the kids' play.

Finally, in approach-avoidance play and games we saw a routine that might be a universal feature of kids' cultures. Here, the threatened kids are always in charge and they collectively produce a routine in which they share the buildup of tension, the excitement of the threat, and the relief and joy of the escape. Furthermore, in approach-avoidance play, kids' social representations of danger, evil, and the unknown are more firmly grasped and controlled. And all of this occurs while kids are playing and having fun, creating their own peer cultures, and preparing themselves for fuller participation in the adult world.

3 "You Wanna Know What Happened Because You're My Best Friend"

....

Making and Being Friends in Kids' Culture

Two girls about four years old, Jenny and Betty, are climbing in a large wooden box in the outside yard of the preschool. Betty has just joined Jenny after playing with another girl, Linda.

"I do like you Jenny, my buddy. I do," says Betty.

"I know it."

"Yeah. But I just ran away from you. You know why?"

"Why?"

"Because I—."

"You wanted to play with Linda?"

"Yeah."

"I ranned away with you. Wasn't that funny?" says Jenny.

"Yes."

"Cause I wanted to know what happened," says Jenny.

"I know you wanted—all the time—you wanna know because you're my best friend," replies Betty.

"Right," says Jenny.

In this example, Betty and Jenny discuss friendship at an abstract level and agree that they are best friends. From the discussion it is apparent that they see each other as best friends because they care about each other. They show an awareness of how their actions affect each other's feelings. This awareness is clearest in Betty's explanation of why she ran away and Jenny's expression of her need to know what happened to Betty.

To most adults, at first glance, Betty and Jenny's talk about their concern for each other's feelings and being best friends might not seem remarkable. However, much research by developmental psychologists on children's acquisition of conceptions of friendship would classify these kids as very precocious in their knowledge about friendship. For example, developmental psychologists use clinical interviews to determine children's friendship knowledge; that is, they ask kids who their best friends are and why or present them with friendship dilemmas (for example, "If a new girl came to your school and your best friend asked her for a sleepover but did not ask you, how would you feel?"). Such studies show that it is not until children are 11 or 12 years old that they see friends "as persons who understand one another, share feelings, secrets and psychological problems." Although Betty and Jenny do not use sophisticated language, their talk and behavior have many of these qualities.

A big reason that developmental psychologists underestimate the friendship knowledge and skills of young children is that they focus on outcomes. That is, they identify and classify children at various stages in the acquisition of adult friendship knowledge in relation to their age or other developmental abilities. There is an assumption here that kids must acquire or internalize adult conceptions of friendship before they can really have complex friendship relations. Surely adult conceptions of friendship are more advanced than those of children and individual children might acquire friendship knowledge and abilities in some general, stage-like manner. Therefore, the work of developmental psychologists on children's acquisition of adult conceptions of friendship is important. However, I am interested in friendship processes in kids' lives and peer cultures. I want to know how preschool kids go about being and having friends. I also believe that it is by engaging in these friendship processes that children acquire more abstract knowledge about friendship. Therefore, I believe it is important to study kids' friendship processes directly in their own worlds and from their perspectives.

PLAYING WITH LOTS OF KIDS: FRIENDSHIP PROCESSES AMONG YOUNGER PRESCHOOL CHILDREN

Let's return to Betty and Jenny. The two girls frequently played together and formed a close relationship in which Jenny was somewhat dependent on Betty. While Betty played with a number of children in addition to Jenny, Jenny had developed a strategy of looking for Betty during free play periods. This strategy frequently worked and the two (often alone but sometimes with others) engaged in a variety of types of play. However, sometimes Jenny's strategy of looking for Betty did not work out so well. Betty might be already playing with some other kids and Jenny's attempt to enter the play was resisted. We saw in Chapter 2 that resistance of entry bids was common among the younger children I studied because the kids tend to protect their interactive space. Sometimes, Jenny did not even attempt to enter a play activity in which Betty was already involved, but rather watched or even followed Betty and other kids around until their play ended.

In fact, that is what happened in the above example. Jenny saw Betty playing with Linda, watched and followed them for some time, but did not try to enter the play. Eventually she went to sit alone on the large wooden block. Betty noticed Jenny but was content in her play with Linda. Therefore, she continued in that play and then joined Jenny when the play was over. Upon joining Jenny, Betty reassured her that they were still "buddies" and Jenny agreed. She told Betty that she followed her around ("ranned away with her") while she played with Linda. Betty suggested that Jenny did so because they are best friends.

Several things are important here. For young children, friends are primarily seen as those kids you are playing with "in the moment." We saw this in Chapter 2, where kids, once they had established a play theme or event, often marked it with the phrase, "We're friends, right?" This reference to affiliation is just that. We're friends because we're playing together, we're sharing, and we're doing it all on our own without the help or interference of adults or other kids.

Most of the three- and four-year-olds whom I studied played with a wide range of other kids (regardless of gender or age). One of the main reasons was that in their experiences in the preschool settings the kids came to realize that interaction is fragile and acceptance into ongoing activities is often difficult. Therefore, the kids concentrate on creating, sharing, and protecting their play. In short, the kids are more concerned with "playing" than "making friends," and, anyway, you make friends by playing with other kids—as many as you can.

Betty was like most of the other kids in the three- to four-year-old group in the Berkeley preschool in that she played with several children regularly. She differed in that she also played with Jenny a lot. Therefore, Jenny became a special or best friend for Betty. However, there can be a downside to best friends. Jenny, unlike Betty, did not play with several other kids on a regular basis and became dependent on her friendship with Betty. As a result, she spent a good bit of time on the sidelines, waiting for Betty or—on the rare occasion—another child to invite her to play.

Among the older group of four- to five-year-olds in the Berkeley preschool, the pattern of most kids playing with several playmates on a regular basis was the same as that of the younger group. But here there was a close group of three boys (Peter, Graham, and Mark). These boys played together frequently and saw themselves as close friends.

They also played on a somewhat regular basis with other kids. Unlike Jenny, the boys were not dependent on one another as friends to gain access to play. They were, however, protective of their friendships and often competitive with one another in their play.

These patterns were especially evident between Peter and Graham. Here's an example.

Peter, Graham, Frank, Lanny, and Antoinette are playing with water in the sandbox in the outside yard. Each child has an individual hose to squirt water into the sand.

"Hey," shouts Lanny, "we made the best waterfall, see?"

"Yeah," agrees Frank.

"That's not a waterfall," says Peter.

"Yes it is," asserts Lanny.

"Lanny's can't. Lanny's isn't," repeats Peter.

"I did the—a waterfall. Right, Frank?" asks Lanny.

"Yeah," says Frank in support.

"Frank's is," says Antoinette.

"Yes. Mine is, isn't it, Frank?" asks Lanny.

"It's mine," says Frank.

"It's both ours, right?" asks Lanny.

"Right," responds Frank. "And we made it ourselves."

"Right," says Lanny.

"Graham, we're not gonna be Frank and Lanny's friends, right?" asks Peter.

"I am," says Graham.

"I'm gonna throw water on you if you don't stop it," says Frank to Peter. "And tell the teachers."

In this example, Peter seems to see Lanny's and Frank's attempt to work together as a threat to his friendship with Graham. He, therefore, suggests to Graham that they not be friends with Frank and Lanny. Graham rejects this suggestion and Frank goes further, saying that he will throw water on Peter and tell the teacher.

In a second example, the children are again playing around the outside sandbox. Two children are between Peter and Graham and Peter tries to get Graham to move over and play next to him.

"Graham, if you play over here when I am, I'll be your friend," says Peter.

"I wanna play over here," responds Graham.

"Then I'm not gonna be your friend," Peter threatens.

"I'm not—I'm not gonna let—," Graham stammers a bit and then continues, "I'm gonna tell my mom to not let you—."

"All right!" Peter interrupts. "I'll come over there."

In both these examples Peter displays insecurity in his friendship with Graham. In the first, this leads Peter to try to pit the two of them against Lanny and Frank in a competition that all the other children reject. In the second, Peter tries directly to control Graham's behavior (get him to play right next to him) by first offering friendship and then threatening to take it away. Graham, who is much more secure in the relationship, admonishes Peter and threatens to tell his mom not to invite Peter to come and play at his house. Facing the loss of playing with Graham at his home, Peter quickly decides to change his behavior rather than keep trying to control Graham.

In these examples and the earlier one with Betty and Jenny, we have somewhat of a contradiction. At an individual level, both Jenny and Peter think about friendship a great deal and try to use friendship to control the behavior of other children. Also, both seem to have a more advanced (or adult-like) knowledge of friendship than most of their classmates. However, as a result of this concern with maintaining their best friends, both Peter and Jenny are not as positively integrated into the peer culture as they could be. While Jenny is often on the outside waiting for Betty, Peter is seen as controlling and manipulative by Graham and his playmates. These examples, especially the ones involving Peter, also remind us that individual children do vary in their social skills and friendship relations. Some children are timid, while others can be bossy, manipulative, or even bullies. Here we see that the complexity of play, friendship, and peer culture is best understood by the direct examination of peer relations in natural settings.

FRIENDSHIP, CLIQUES, AND GENDER RELATIONS AMONG OLDER PRESCHOOL CHILDREN

While younger preschoolers' peer and friendship relations are closely tied to establishing and maintaining play, older preschoolers whom I studied were more confident in their social skills and more reflective about their peer relations and friendships. The groups of five- and six-year-olds I studied in Berkeley, Bloomington, Indianapolis, Bologna, and Modena all displayed (1) more reflective awareness and talk about friends and friendship; (2) more differentiation in their peer cultures and the emergence of subgroups or cliques; and (3) a good deal of gender separation in their play. However, like the younger children's friendships, these patterns varied across the cultural and subcultural groups and particular preschool settings.

Friendship, Cliques, and Gender in Upper-Middle-Class American Preschools

In the private upper-middle-class preschools I studied in Bloomington, Indiana, all the kids played together, but over the course of the school year gender-based cliques developed. By cliques I mean groups of kids who played together on a regular basis and referred to each other as good or best friends. Members of these cliques did not so much reject or refuse to play with other kids, but rather tended to seek each other out for particular types of play and talk. Cliques of boys preferred sports and run and chase games, while girls enjoyed play with dolls, toy animals, and role-play. Although boys often marked their close friendships in their cliques, as did the girls, the boys were more open to nonclique members and were less likely to get into disputes about friendships within the clique than the girls. When disputes did arise in boys' cliques, they were usually over the nature of play (choice of a game or disputes about game rules), and while they could be intense or even aggressive, they were normally short lived. Disputes and conflicts in the girls' cliques, on the other hand, were more frequent, emotionally intense, and long-lasting. Sometimes girls stayed mad at each other for several days, but in the end made up and were best friends again.

Let's consider an example of conflict within a girls' clique among five-year-olds in one of the Bloomington preschools. Actually, there were two overlapping cliques with three girls in each clique, but with the six girls (Megan, Shirley, Mary, Veronica, Vickie, and Peggy) all playing together often. In one of the cliques made up of Megan, Shirley, and Peggy there was a good bit of competition among the three about who was the leader and about the strength of their friendships. In the following example the competition between two of these children (Megan and Shirley) is apparent as Shirley resents the fact that Megan will not accept her into a play theme she has organized with Veronica and Mary.

In the outside yard of the preschool Mary and Veronica are pretending to be pet ponies that belong to Megan. Megan, who devised this play theme, has two pom-poms that she uses to direct the ponies' behavior. Shirley, who has been playing elsewhere, sees Megan and the others and comes over and asks to play. Megan at first ignores Shirley and then says she can't play. Megan, Mary, and Veronica now move to another area of the yard, and Shirley follows and again asks to play. But Megan says she cannot.

After two more unsuccessful attempts, Shirley asks Mary and Veronica to be her ponies and to abandon Megan. Mary and Veronica refuse and Megan tells Shirley, "I said you cannot play!"

Mary and Veronica do not actually reject Shirley, but they obey Megan and continue to take the role of her baby ponies. This gives them a bit of freedom, because they can go off to play by themselves but still be Megan's pets and return to her now and then.

Shirley persists in trying to persuade Megan to let her play, but without success. Shirley now starts to cry and tells Megan, "You're hurting my feelings!" Both girls now exchange threats about not being "best buddies" anymore and not getting invited to birthday parties. Shirley also threatens Mary and Veronica, but they stick with Megan. Eventually, Shirley goes and tells a teacher that Megan "is being mean and won't let me play." The teacher suggests that Shirley should play with someone else.

Now very upset, Shirley goes over and shoves Megan and Megan shoves back. Before long both girls are crying. It is now time to go inside and when we get in the classroom the teacher sits the two girls down and talks about the problem. However, the girls refuse to make up and sit at opposite sides of the room during snack. When I return later in the day after nap time, I see Shirley and Megan sitting together and watching a video about a circus with the rest of the kids. The two girls are holding hands and actually kiss at one point.

Emotional disputes among girls (and less frequently boys) who considered themselves close friends were frequent in this and other preschools I studied. In fact, one Italian teacher remarked to me after settling such a dispute, "These best friends are fighting all the time!" A possible reason for this pattern is that best friends in preschool are in frequent contact, expect a great deal from one another, and are often insecure about maintaining close friendships. In the case of Megan and Shirley, it seems that Shirley was upset not only by Megan's rejection but also by the possibility that Megan might now prefer Mary and Veronica to her. Shirley's claim of "hurt feelings" and her denial-of-friendship threats were unsuccessful in the short run. However, Shirley and Megan did make up shortly after their fight, and it could be argued that their dispute actually strengthened their relationship because it forced them to think more about its importance in their lives.

As we have seen in the previous examples, friendship processes and gender relations are often related in the peer culture of older preschool kids. Although there was cross-sex interaction in structured activities and during meals in the Bloomington preschools, it was rare during the children's free play time. This finding of gender segregation in free play among children five to six years old is common in research on children's play and is often tied to differences in styles and types of play (for example, boys preferring more physical, rough-and-tumble play and girls more relational role-play). However, these findings of gender separation in young children's play hold more strongly for middle-class white American children than for other ethnic, racial, and cultural groups. As we will see later, I found less gender segregation among the Head Start and Italian children than among the middle-class white American children.

When girls and boys did play together during free play in the Bloomington preschools, their play activities often took the form of what the sociologist Barrie Thorne terms "borderwork" and defines as *activities that mark and strengthen boundaries between girls and boys.* Borderwork, although collectively produced by girls and boys, builds a heightened awareness of gender differences. Here's an example from my field notes in one of the Bloomington preschools.

Anita, Ruth, and Sarah are chasing Sean and David, who are pretending to be afraid of the girls. At one point the boys come over to a big rock where I am sitting and claim the rock as home base. Anita and Sarah run over and pull up their shirts and say, "You want to see my bra?"

"I have a bra for my belly button," adds Anita as she pulls up her shirt to show her belly button.

Later, when the boys have run off, Anita tells me, "I really have a bra at home!"

Although the girls are far too young to actually have breasts, they are aware that females develop breasts and wear bras. Furthermore, they seem to grasp that displaying breasts is threatening for boys in some way and they use this knowledge to enhance their run and chase play. While the children at the Bloomington preschool were by no means completely gender segregated, this type of borderwork in cross-sex play promoted a charged atmosphere that made boy-girl play risky.

Talk about sex and gender relations also arose in discussions that occurred in structured settings like snack time where a mix of girls and boys was common.

During snack, Veronica says that she and Martin are going to get married.

"Yeah," Martin agrees, "and live in New York."

"Are you going to kiss and do sex?" asks Mark.

All the other kids laugh at this, and shortly after, meeting time is announced.

Kids are often intrigued by what they have heard about adults and their lives and occasionally use this information to predict what their own life courses might be like. In this episode, discussion leaps from talk about Veronica and Martin's idealized future to joking about kissing and sex. It is not clear how much the kids actually know about taboo subjects such as sex. However, the information they do have is likely to be passed on to them from their peers or gleaned from mass media, and, as is the case here, the kids like to demonstrate their competence in this area to their peers. What is most interesting is that the talk occurs where boys and girls are put together by adults, and where they feel relaxed enough to pursue and even joke about such topics without being teased or ridiculed.

Friendship, Cliques, and Gender in the Indianapolis Head Start Center

Although there was a good bit of gender separation in the free play of the Head Start kids, there was more cross-gender play than in the

Bloomington preschools. Also although some kids played with certain kids more than others, there were no strong cliques at Head Start. This lack of cliques could have been influenced by the fact that the children spent much less time together during the school term because of the program's limited resources. The kids were together for only about three hours a day, four days a week, compared to seven to eight hours a day and five-day programs in the Bloomington preschools. The lack of cliques might also have been the result of the strong collective ethos of the Head Start program, which stresses that *every child is important in the group.* In any case, although conflict and disputes occurred among the kids, they were seldom related to disputes over or between friends.

Both cross- and same-sex play themes at the Head Start center were interesting because of their variety and their differences from gender stereotypes. The boys were not hesitant to engage in family role-play either with girls or on their own. On several occasions, groups of four or five boys entered the family play area, took out dishes, set the table, and pretended to prepare and eat meals. They also enjoyed sweeping the floor, getting the house in order, and making phone calls. In addition to this nonstereotypical role-play by the boys, a group of boys once transformed the family role-play into a barbershop.

Four boys (Charles, Jeremiah, Antwaan, and Joseph) enter the family play area and begin setting the table where I am sitting. Antwaan finds a comb and begins to comb my hair. The other boys now stop setting the table and gather around us. Charles picks up a toy camera and tells me to say "cheese" and takes a picture as Antwaan continues to comb my hair.

"This is a barbershop," says Charles. He then picks up a plastic knife and a tissue and begins cutting my hair and trimming my beard. He holds the knife and tissue as if he is giving me a "razor cut." Antwaan has now stopped combing my hair and has taken a broom and begun sweeping the pretend hair from under my chair.

Jeremiah now also gets a knife and trims my beard while Charles continues to cut my hair. "Hey, watch out," Charles tells Jeremiah. "I have to finish him up because he has to go to a concert." Jeremiah backs off, and Charles picks up a whisk broom. "I'm brushing him off," says Charles.

Charles finishes brushing me off and puts down the whisk broom to make a phone call, using the toy phone near the pretend barber chair. Meanwhile, Jeremiah is trimming my beard again and Joseph picks up the whisk broom to brush me off some more. Antwaan now takes the big broom he is using to sweep the floor to brush off my pants and shoes.

After making several phone calls, Charles returns and says, "Bill better be finished because his girlfriend is going to get him." I assume that means pick me up for the concert. In the meantime, Jeremiah and Joseph begin to struggle over the use of a comb, and Jeremiah says, "I am the barbershop man!" I suggest that Joseph can shine my shoes and he accepts this role while Jeremiah continues to work on my haircut.

I ask Jeremiah how much my haircut will cost and he says, "Five dollars." "No, it costs ten dollars," says Charles." I say that five dollars sounds good to me and pretend to pay Jeremiah. Now Charles and Joseph turn a mirror in the play area so that I can see my haircut. About this time, the teacher tells the boys to begin cleaning up the area and the play ends.

A barbershop is clearly a male domain and a place for a high level of sociability among males in the African-American community where the children lived. Not only do the boys pretend to cut my hair in a realistic fashion with the clever use of the knife, comb, and tissue, there is also a discussion of why I need a haircut. I am going to a concert and my girlfriend is picking me up. The boys also make a number of phone calls, which is a typical activity in barbershops in their community. Overall, they transform what is normally a family play area and activity (making meals and cleaning house) into a male occupation and setting (barbershop) that they have no doubt experienced quite often in their young lives.

The girls enjoyed family role-play and other activities like arts and crafts that the boys seldom engaged in without prodding from the teachers. However, the girls also relished competing with and challenging the boys. Such competition often occurred in the gym and took the form of borderwork that we discussed earlier.

As we wait in line to go to the gym, several girls tell me that there is a girls' clubhouse in the gym and the boys are not allowed in it. When we get in the gym, the teachers engage the kids in an activity in which they form a circle and are thrown a large red ball in a random order. The activity is to help build motor skills and the kids enjoy it, especially trying to guess when the ball will come their way. Right after we finish playing ball, several girls run to a climbing house. When I arrive I hear them say, "This is the girls' clubhouse!" They chase out two boys who resist at first, but then run off. There are seven girls involved in the play in the clubhouse, including two from another class now in the gym. It is clear that there is a history to this competition over the clubhouse and girls clearly relish running the boys off. Later the girls abandon the clubhouse and several boys enter, climb, and yell, "Now we got the clubhouse!" The girls are playing elsewhere and do not notice this invasion.

The conscious labeling of gender and negotiation of space demonstrates that gender is an important marker of identity among the Head Start kids. In this case, gender is a salient enough identity that two girls from other classes join in the play and are immediately accepted because they meet the gender requirement for being in the "clubhouse." However, as discussed earlier, this type of borderwork heightens gender differences and the girls and boys actually work together to create the competition. Thus, it is one way for girls to play with boys while at the same time using the structure of the play to claim that such cross-gender play is unwanted. In short, the girls (who almost always instigated the borderwork) are saying you can't play here, but we dare you to try. The fun of the play is taunting the boys and running them off, which would not occur if the boys ignored them. It takes two to tango, and play about not wanting to play with boys is actually play with boys—but on the girls' terms.

The assertiveness of the girls in the Head Start program was also apparent in certain personal interactions and relationships. Several of the girls in the class were very active in teasing with boys and other girls. One girl, Delia, frequently stood up to boys and relished taking them on in verbal disputes.

Delia asks to print her name in my notebook when she sees me taking notes in jail (where I have been locked up by several boys playing police). I hand her the notebook and pen, but as she prints, Dominic comes over and says, "Give me that notebook." Delia tells him, "Get out of my face while I write this name!"

"You're talking to the police," I remind Delia.

Delia then says, "Get out of my face police!" She finishes printing her name, hands me back the notebook and pen, and walks off.

Delia's assertiveness can also be seen in her relationship with Ramone, who has a crush on her. He told several of the other children and me that Delia was his girlfriend and that he visited her at her house. Delia denies both claims. Still, Ramone does not give up and continues his pursuit of Delia.

Alysha and Delia are putting together a large puzzle of a school bus on the floor near the circle area. They are working together to fit the pieces properly. Because the puzzle has such large pieces, it is not demanding and the girls make quick progress. When they are about to finish the puzzle, Ramone comes over and asks to play. Delia says, "If you play with girls, then you are a tomgirl!" Ramone takes this as a rejection and moves away briefly, but then comes back and picks up a piece of the puzzle. Delia takes it away and says that when she plays with boys she is called a tomboy, so if Ramone plays with them he is a tomgirl. She also says that they do not want Ramone to play anyway and Alysha agrees. Ramone now moves to another part of the classroom.

What is intriguing about this exchange is Delia's use of the term "tomgirl" to discourage Ramone from playing with the girls. Not only does this term seem to be an interesting adaptation of the "tomboy" label, it also changes the nature of the rejection from "Don't play with me" to "Don't play with girls." Furthermore, when Delia explains the novel term "tomgirl," she reveals that she herself has been teased for playing with boys. Overall, from these examples, we see the complexity of the Head Start children's construction and use of gender in their peer relations.

Friendship, Cliques, and Gender in the Italian Preschools

In Italy I became part of kids' peer culture in preschools in Bologna and Modena. In both schools there was a strong emphasis on communal and collective values, and in Bologna the teachers encouraged the older kids who had been in the school for one or two years to work and play with the new three-year-olds. As a result, there was a great deal of both cross-age and cross-gender play in the peer culture. Still, the five- to six-year-olds in the school also formed close friendship cliques that were primarily gender segregated. Many of the kids in these cliques frequently played together in school and also visited each others' homes. Toward the end of the school year the older kids began to realize that their lives would change when they went to elementary school, and that they might not be able to maintain their close friendships. This uncertainty sometimes led to debates and disputes. One such dispute occurred among three close friends (Mario, Enzo, and Dante), who were all about three months shy of being six years old. The three were good friends, but there was a history of competition between Enzo and Dante over Mario's friendship. The boys had been playing a board game and were now considering alternatives. Dante has suggested that they play with building materials called "Clipo" (grooved plastic objects that clip together), which he liked to use to make spaceships. Enzo was not happy with this idea.

"Yes, but why do we have to do everything you do, Dante?" Enzo complains.

"But no, it not right Enzo," says Dante. "One day I heard something from Mario that you want to construct the best things and that you always exclude us."

"On the contrary," protests Mario, "it's not true. I didn't say that he always wants to exclude us."

"With Clipo I always built what came to me," says Enzo, "and then afterwards one day you started claiming that I built something better. I remember it. If you don't remember it, I do. Understood Dante?"

"Listen Enzo, yes," replies Dante, "but I told you that it's something that Mario said. I didn't want to believe it—."

"But excuse me," interrupts Enzo, "Mario didn't say it because I was there that day."

"I know it, but you were somewhere else," says Dante. "Mario told me secretly, and you could not—."

"No, he did not tell it secretly," interrupts Enzo, "because if he told it secretly, I would not be his friend anymore."

"But you were there, but you were somewhere else [in the school]," Dante responds.

"But where was I?" demands Enzo.

"You were—do you know the garden down there?" says Dante. "You were there on the other side of the little swimming pool playing with others while Mario was behind the tree there and told me that 'we're back here because he wanted to exclude us from the game to obtain—.'"

"What?" interrupts Enzo incredulously.

"I did not say that," Mario declares.

As I noted, Dante and Enzo often competed for Mario's friendship. Therefore, Enzo's opening remark can be interpreted as something more than just opposing play with Clipo. It can be seen as a move in a strategy to build solidarity with Mario (stressing their shared irritation with Dante's constant choice of Clipo) and build a wedge between Dante and Mario. The fact that Dante goes beyond a simple denial of Enzo's claim (that they always play with Clipo) is important. Not stopping with the denial, Dante reports an event in which Mario shared with him a negative evaluation of Enzo.

Mario now enters the discussion denying Dante's claim that he said Enzo did not want to play with them. Surprisingly, Enzo ignores Mario's denial and quickly responds to what he sees as Dante's implicit criticism of him that he (Enzo's) constructions are better because they are copies rather than originals. Dante, however, backs away from Enzo's challenge by reminding him that he was simply repeating something that Mario had said. Rather than accept Mario's earlier denial, Enzo tries to discredit the validity of Dante's report of the past event ("I was there that day"). At this point, proving Dante wrong seems more important to Enzo than accepting Mario's implicit support of his side of the debate.

The complexity of the talk and strategies of Dante and Enzo is highly impressive, given that the boys are only six years old. Also, we see that disputes about friendship, once under way, can move in unplanned directions. Mario, for example, now finds himself caught in the middle of a dispute that holds little interest for him.

Things get even more difficult for Mario with Dante's assertion that Mario told him something about Enzo secretly. I found that it was not uncommon for children—in both Italy and the United States—who had close friendships to share secrets, such as plans to visit each other at home, sit together at lunch, or ask the teacher to work on a project together. Overall, the secrets were of little import beyond marking the relationship as special. In a few cases, however, secrets involved revealing negative evaluations of others as a way of building solidarity in the friendship group and placing one's own group above others.

Given this shared significance of secrets, Enzo understandably reacts very negatively to Dante's claim of secrecy with Mario. One does not say negative things in secret about a friend to others. At this point, Enzo is not only disputing Dante's version of the reported event but also implicitly threatening Mario ("I would not be his friend anymore") if Dante's version turns out to be correct. Mario denies that he told Dante something in secret. However, Enzo and Dante continue their dispute.

"I didn't want to exclude you from the game," says Enzo to Dante.

"I told you that—," says Mario to Dante.

"What did you tell me? Tell me!" demands Dante.

"I did not say—," says Mario.

"No," Dante interrupts Mario again. "It's not true Mario."

"Listen, but really Enzo—" starts Dante.

"Do you know," interrupts Mario, "that Mario is the name of a Roman warrior? It is a very old name used by the Romans."

"Eh, but I wanted to tell you something else Enzo," says Dante ignoring Mario. "I don't watch robot cartoons and then decide to copy the spaceships. I invent them and I can do them by myself."

"But this is because you have all the robots at home," says Enzo. "I saw you when I came to your place, at your party."

Sensing the seriousness of Enzo's concern about secrets, Mario tries to change the topic. Although unsuccessful, Mario demonstrates impressive interpersonal skills. By subtly attempting to move the talk in a new direction, one that he believes might interest his peers (the origin of his name being that of a Roman warrior), Mario hopes to draw the attention of his quarreling playmates away from their struggle. If successful, he could also extricate himself from the middle of their feuding. Dante and Enzo, however, ignore Mario's strategy and continue to argue about the originality of Dante's constructions with Clipo.

This example demonstrates the highly integrated nature of the older Italian children's friendships. It also reminds us that to appreciate this complexity we must be aware of the historical and contextual features of friendship in children's peer cultures. Children who share long histories of interaction in small, cohesive groups often develop friendship skills that can be captured only by joining and becoming part of these groups.

In Modena the group of children I joined in the middle of their third year together had created a highly communal and rich peer culture. All the children knew each other well and most considered themselves good or best friends regardless of age or gender, and there were no exclusive cliques. Several other factors also contributed to the highly communal peer culture. One was the school curriculum. Although some activities were clearly teacher directed and some free peer choice, many others had features of both. Most days after group meeting time (teacher-directed) and outside meeting time (free play), two to four children usually worked on art or literacy projects with teachers, several other kids tended to other aspects of the projects (drawing or painting pictures, cutting paper, and so on) without teacher supervision, and still others selected free play activities. The kids often rotated seamlessly in and out of structured and semistructured activities and free play. Because structured activities were normally gender mixed, so too, were many of the semistructured activities and free play episodes I observed.

Another factor that contributed to the lack of differentiation in the peer culture was the popularity of certain play routines. While the kids participated in traditional gender-type activities like physical play and games (riding bikes, soccer, and superhero play) for boys and playing with dolls for girls, another typical gender-typed activity, dramatic role-play, had a more complex pattern. Although mainly girls engaged in domestic role-play, both boys and girls often participated in types of role-play that blurred and stretched gender stereotypes.

The most common was animal-family role-play where both boys and girls pretended to be wild dogs, lions, or tigers. However, in addition to this type of role-play, the Italian children often re-created variety or game show television programs, which are very popular in Italy. One program that was particularly popular at the time of my research in Modena had two central characters, a male host and a woman gypsy fortuneteller, along with two couples (usually married) who competed for prize money. The show also featured elaborate sets and singers and dancers in colorful costumes. In their play the kids focused primarily on a part of the game portion of the show where each of the two couples selects a card from a set of seven laid out by the gypsy. The object was to answer correctly questions associated with various cards and win money, while avoiding the *Luna Nera* ("Black Moon") card, which, if selected, eliminated them from the contest and lost them any money previously won. All of this occurred within a choreographed routine in which the gypsy flipped the cards with suspenseful flair, all to a catchy musical refrain.

The show was very popular among all the kids at the school, but one boy, Dario, especially liked it. He frequently organized other boys and girls to play the game using regular playing cards with the ace of spades representing the *Luna Nera*. He also sometimes brought a toy version of the television game show from home to school. Perhaps because the gypsy was the central character, both boys and girls wanted to repeat the game several times with each having a turn in this desired role. Thus, the kids shared the fun of the reproduction of the show in mixed-gender play without the girls and boys embodying only its obvious gender-typed features (a boy as the host and a girl as the gypsy).

The Modenese kids, like other Italian preschool kids I studied, valued verbal debate and discussion. Debate of this type, or what Italians refer to as *discussione*, was an integral part of the school and peer culture. However, the kids not only used debate to build a collective ethos in the school, they were also very sensitive to instances where debates escalated into personal conflict. In these cases the children worked collectively to ease tension and restore harmonious relations in the group.

One way the children did this was through humor. Often in verbal debates a child who might have been losing ground accused the other of being a "know-it-all." Such name-calling sometimes escalated to more serious conflict. However, escalation was often quashed by other kids (not directly involved in the debate) who supported the offended party with humorous remarks. A favorite was to refer to the party with the upper hand as *"professore"* ("professor"), a backhanded compliment implying that the agitator is taking on airs. Here's an example. Valerio falls and is crying. Sandra says it's his own fault and now he has hurt his foot because he was running around too much. This diagnosis is taken as insulting by Valerio, who is now injured and mad. Viviana, standing nearby, observes, *"Ah, Sandra adesso è dottoressa"* ("Ah, Sandra is now a doctor"). Both Sandra and Valerio laugh at this remark, ending the conflict and also, it seems, the pain in Valerio's foot.

In some instances, humor was not enough to quell conflict and serious disputes occurred. In many such cases, however, uninvolved kids often negotiated peace between the warring parties. Here's an example from my field notes.

Carlotta and Sofia get into a dispute over whose turn it is to ride an available bicycle. There is some pushing and shoving and Carlotta stalks off very angrily. I had noticed these two getting mad at each other before. I now see that Elisa is bringing Sofia over to Carlotta, so I follow close behind. Elisa tells Sofia and Carlotta to stay alone and work it out. Carlotta is quite upset and begins to cry. Stefania, Federica, and Elisa now come over to Carlotta and Sofia as does Marina. Elisa tells Marina to take Sofia aside and talk to her, because she (Elisa) will talk to Carlotta. Sofia begins to cry and is comforted by Marina and then by Elisa and Stefania. Marina takes Sofia to the teacher briefly, and then the two go to get Elisa, who is with Carlotta.

Meanwhile, Renato comes over and talks with Carlotta and Elisa. Marina brings Sofia over. Marina makes a joke and everybody laughs. But Carlotta and Sofia are still upset, and Sofia says that Carlotta is a "big liar." The others try hard to overcome this problem. Eventually, the two seem to agree not to fight anymore, but they have not made up. Later when the children go inside, wash up, and sit down waiting to go to lunch, I notice that Carlotta and Sofia have made up and are sitting next to each other. They are very happy and laughing. They are also glad when Marina (who is one of the waiters for lunch) selects the two of them for her table. They run off with their tablemates with hands on shoulders.

This was one of several examples where a small group of children (usually four or five) worked together to settle a serious rift between two of their playmates. In this instance and in almost all of these cases, a teacher or teachers became aware of the problem but left the children alone to settle things themselves. The children saw serious conflict between or among their peers as a threat to the strong group identity of the peer culture and worked collaboratively to reduce this threat.

We can contrast this example with an early one we discussed in the Bloomington preschool. There, when two girls, Megan and Shirley, got into a serious dispute, other children (even the two girls playing with Megan when Shirley tried to enter the play) stayed out of the dispute. In short, the dispute was seen as a private matter and somewhat external to the group. The teacher worked with the two middle-class American girls to get them to talk things over and eventually they made up. However, in the Modena preschool, as we have seen, disputes are not viewed as private matters but as threats to the group as a whole. Overall, the nature of conflict and the way it was handled in the kids' friendships and peer relations in Modena again demonstrate the strong social cohesion of this group of children.

FRIENDSHIP AS SITUATED KNOWLEDGE

Earlier we discussed how developmental psychologists have been interested mainly in children's knowledge of friendship as an abstract concept or a set of skills that can be described and evaluated separately from the social contexts in which they develop and or used. This approach is useful for charting the individual child's acquisition of friendship knowledge and skills over the course of childhood. However, it tells us much less about how kids go about making and being friends and how friendship processes are situated in children's everyday lives.

When we say friendship knowledge is situated we mean that it, like all social knowledge, develops from social action (doing things with others) in a variety of types of social settings over historical periods. The white American, African-American, and Italian preschoolers all saw other kids whom they played with as their friends. Friends are kids you do things with. The older kids in the various preschools we discussed also had some insight into the notion of best friends—kids you have a special relationship with, whom you care about and share secrets with. However, knowing these two general facts about the kids' friendship knowledge only scratched the surface of the complexity and variation of friendship in their peer cultures. To appreciate this com plexity we had *to take seriously the social situations in which friendship knowledge and skills develop.* When we did so, we saw how the gender structure and size of the group, the amount of time a group of kids share their lives, the nature of the preschool curriculum, and the social and cultural values of the group and of the wider society are all related to kids' friendships. However, it is not easy or advisable to try to pull these factors from their social moorings and try to measure in some way how they affect kids' friendship. Instead, we must embrace the very situated nature of friendship, make ourselves part of those situations, and see and feel and try our best *to understand what kids' friendships are like while they are kids and during their childhoods.*

4

"You Can't Talk If You're Dead"

· ·

Fantasy and Pretend Play

Joseph and Roger are building with small blocks at one of the work-tables in the Berkeley preschool. I am sitting with them, watching. I notice that Joseph's building is getting very tall. A teacher, Catherine, also notices and comes over to the table.

"Boy, what a tall building!" Catherine says.

"Yeah," says Joseph. "It's the Vampire State Building!"

Catherine and I look at each other and laugh. Catherine now moves away, and I look back at Joseph and Roger. They're not laughing. They continue to work on their buildings. Here the boys have mixed together two aspects of their pretend world that are very important to them: tall buildings and monsters. In the process they have misnamed the Empire State Building something that seemed funny to us adults. But for the kids the name seemed logical and correct.

When it comes to pretend play, make believe, and fantasy, kids do not just have a different perspective than adults; they are highly skilled producers and directors of their own imaginary worlds. In fact, I believe that young children (three- to five-years-old) are more skilled at creating, sharing, and enjoying fantasy play than are most older children and adults. To gather and interpret evidence to support this claim, it is necessary to appreciate kids as *active* consumers and producers of their own symbolic culture. By carefully observing and videotaping numerous examples of children's fantasy play, I gained an understanding of how children produced it. I also discovered that kids take basic themes and aspects of adult-produced literature, movies, music, and television and then use and embellish them in spontaneous fantasy play in their peer culture. Many of these spontaneous and improvised performances address important socio-emotional needs in early childhood.

SPONTANEOUS FANTASY:
WHAT IT IS AND HOW IT GETS PRODUCED

Almost all definitions of play include some reference to fantasy and the absence of rules or strict guidelines that structure the activity. In spontaneous fantasy, children become animals, monsters, pirates, train engineers, construction workers, and so on and structure the activity as it emerges. They often do this by manipulating objects like toy animals, building blocks and other construction materials, toy cars, trains, and the like. I want to distinguish spontaneous fantasy, which I am defining in a very general way, from socio-dramatic play (more about that in Chapter 5), in which children take on or embody roles that exist in society (like mothers, fathers, or various occupational roles).

In spontaneous fantasy there might be mothers or fathers, firefighters, soldiers, and race car drivers (roles that exist in society), but the children animate objects that represent these figures rather than embody them. In preschools, spontaneous fantasy often occurs around sandboxes or tables, in building and construction areas, and sometimes at worktables as part of or as transformations of artistic or literacy projects. The expectations kids bring into these areas are not well defined. They know they will play with certain objects (toy animals, blocks, cars, and so on), but they seldom enter the areas with specific plans of action. The play activity emerges in the process of verbal negotiation; shared knowledge of the adult world, although referred to at times, is not relied upon continuously to structure the activity. In short, the activity is highly creative and improvised.

In spontaneous fantasy, children use a number of identifiable communicative strategies. Here are two short sequences from a longer play event that has a general theme of danger-rescue produced by children in the Berkeley preschool. Later in this chapter I will return to these two play events and examine the substance of this and other themes (lost-found and death-rebirth) in spontaneous events.

Rita, Leah, and Charles (all about four years old) are kneeling around a sandbox playing with toy animals. We begin videotaping the play shortly after the children enter the area.

"Help! Help! I'm in the forest," says Rita as she moves a toy horse up and down in a hopping fashion.

Charles hops his rabbit near the center of the sandbox and says, "After you Madam, into this fence."

Leah then places a goat next to Rita's horse and asks, "Where's your home?"

"Into this sandpile," says Charles as he moves his rabbit under the sand.

"In here!" shouts Leah, moving her goat to the top of the sandpile.

Charles removes his rabbit from the sand and places it near Leah's goat, "Into this—."

"Into this hole!" yells Rita cutting Charles off as she moves her horse near his rabbit. Then Leah and Rita put their animals in the sand and cover them up. Rita hums "Do-do-da" as they do this. Charles watches with rabbit in hand near the top of the sandpile.

Highly important in spontaneous fantasy is the children's use of paralinguistic cues like voice quality and pitch. In their talk, they use high pitch, heavy stress at the end of utterances, and rising intonation to mark that *they are the animals they are manipulating*. The children begin to structure their play through their manipulation of the animals, calls for help, and identification of a home inside the sandpile.

However, the play is just beginning to emerge. No suggestions are offered about a plan for exactly what the play might involve (for example, "Let's pretend we are the animals and there is a big storm"). Instead, the kids rely on the nature of their speech and actions and the responses of their playmates to signify that they are playing together and must fit into the fantasy play with appropriate responses when necessary. In this case, appropriateness is tied to the ongoing play and is spontaneous in that the kids build the play by plugging into and expanding on each other's contributions.

The play continues as Charles moves his rabbit up and down on the top of the sandpile and says, "This is our b-i-i-i-g home! And I— I'm a freezing squirrel." He then buries his animal in the sandpile. Leah takes another animal and buries it in the sandpile.

Rita then takes her horse from the sand and says, "And this got out. And I'm freezing! Whoop-whoop-whoop-whoop!" Rita moves her horse up and down with each whoop.

Leah is smoothing the sand so that the pile is higher and she says to Rita, "Get in the house."

Rita now puts her horse in the sand and covers it saying, "Oh— wow—get in the house!"

Rita next picks up a handful of sand and sprinkles it onto the pile. As she does this she shouts, "Oh look, it's raining. Gonna rain."

Charles now takes his squirrel from the sand and says, "Rain. It's gonna be a rainstorm!"

"Yeah," replies Rita.

"And lightning. Help!" yells Charles.

Charles now moves his squirrel away from the sandpile to the other side of the sandbox and says, "But I won't be hit, though—cause lightning only hits a bigger—bigger—will hit our house cause it's the biggest thing. Cause our house is made of—."

Leah interrupts Charles as she and Rita take their animals from the sand saying, "Going-going."

"But our house is made of steel," continues Charles. "So the lightning just fall to the ground." Charles now returns his squirrel to the sandpile.

"Right," says Rita. "Won't get horsie." Rita and Leah now place their animals back in the sandpile.

Charles introduces the idea that the sandpile can be a home for the animals and that he (the rabbit he is holding) is a freezing squirrel. In his first speech in this sequence, Charles stresses the adjective "b-i-i-i-g," elongating it to mark his transformation of the sandpile into a home for the animals. Rita then takes her horse from the sand and connects her activity to Charles's by repetition of the phrase, "I'm freezing." Here Rita is tying her action to Charles's earlier one by repeating his original idea (the animals are freezing when outside the house). Leah then takes a turn telling Rita to "get in the house." Here we have an expansion on Charles's original notion in that Leah is telling Rita that it will be warmer inside. Rita responds appropriately by putting her horse back in the sand.

The interesting thing about Rita's action is that her speech and physical manipulation of the toy horse are fused: She describes her action as she does it. I have found in spontaneous play that kids consistently provide verbal descriptions of their behavior. When viewed from an adult perspective, such descriptions might be labeled as "egocentric" speech. The psychologist Jean Piaget characterized much of the language and thought of preschool children as egocentric, arguing that it was basically emotional and self-directed rather than social. However, the description of ongoing activity in spontaneous fantasy is important in that it cues other participants to what is currently occurring and allows the kids to take up and expand on the emerging social event. As we saw, that is what happened in this case as the kids responded to descriptions of actions to extend their play.

After Rita puts her horse in the sandpile home, she does something that is an excellent example of why I term this activity "spontaneous fantasy." As she covers her horse with sand she notices that the sand is falling on the pile like raindrops and says, "Oh, look it's raining. Gonna rain." The emergence of the rain was spontaneous and unpredictable. It occurred because Rita happened to be sprinkling the sand from above rather than raking it onto the pile. In her utterance, Rita calls attention to her spontaneous extension of the play and then states it. Thus, she provides for the organization of her behavior and a semantic base on which the other children can build.

This is just what Charles does, first marking Rita's new addition and extending it to a "rainstorm" and receiving confirmation from Rita. Charles then goes on to add "lightning" to the rainstorm and then suggests leaving the house to avoid a lightning strike because the house is the biggest thing.

As Charles communicates this idea he takes his animal from the house and so do the two girls. However, in this very process of cautioning about the possibility of lightning striking the house, Charles reverses his thinking. He decides that the house is made of steel, so "the lightning just fall to the ground." He is describing a sort of "lightning rod" idea. He then puts his animal back in the sandpile, and so do the two girls, with Rita noting that the lightning "won't get horsie."

An awful lot has happened in this sequence in just a few minutes of play. The children reach an agreement that they are the animals they animate, that the animals have a home, that it is cold outside the home and warm inside, that it begins to rain, that the rain becomes a storm with lightning, that the lightning might hit the house because it is a big target, and finally, that the house is safe from a lightning strike because it is made of steel. The kids accomplish the collaborative fantasy play through subtle use of various features of language. In no case do they offer up a script or plan of action nor do they use stage directions that place them outside the action (for example, "Let's pretend there's a rainstorm"). In short, the fantasy play and the beginning of a danger-rescue sequence (discussed later) is constituted totally in the social interaction itself. This complex, improvised feat is accomplished by the use of paralinguistic cues (voice, pitch, intonation), orchestrated manipulation of play objects (the toy animals and sand), verbal descriptions of actions, repetition of speech and action, and semantic tying and expansion (for example, from rain, to a rainstorm, to lightning).

It is easy to overlook the complexity of this type of spontaneous play, because for most adults it is seen as "just kids playing make believe." However, I challenge any adult to try to produce such make-believe play in this totally in-frame (that is, without out-of-frame discussion of plans for action), implicit, improvised way. It seems easy until you analyze it very closely or try to do it. In fact, adults appreciate the complexity of such improvisation *if adults are doing it*. We pay to see improvised comedy shows like *Second City* and we sing the praises of, and lavishly reward economically, comedians like Robin Williams. However, for preschool children, it's just "kids playing make believe around a sandbox." We see what we look for.

DANGER, BEING LOST, AND DEATH-REBIRTH: THEMES IN SPONTANEOUS FANTASY

Although much of children's fantasy play is spontaneous and improvised, children's shared knowledge of aspects of the adult world and their own peer cultures is important for its production. Shared knowledge is important at the micro, turn-by-turn, level of fantasy play (for example, knowing that rainstorms often contain lightning) and for underlying themes or plots on which specific play sequences are built.

As we saw in the last example, an underlying theme of danger-rescue developed in the kids' play with the toy animals. Two other themes I have discovered in children's play are lost-found and death-rebirth. All of these themes are similar in that they share the general frame of a buildup and release of tension. In this way they are similar to the plots or arches of stories or narratives in general (for example, fairy tales and films for children). Themes are not, however, scripts or plans. They are not that specific and are much more malleable. Thus, children rely on implicit, shared knowledge of things like danger, death, and being lost, but have ample latitude in generating detailed fantasy action in line with these themes.

Danger-Rescue Theme

The kids' ability to create danger seemed almost limitless. There were rainstorms, fires, tidal waves, snowstorms, falls from cliffs, threatening animals, earthquakes, quicksand, and poison, to name just a few themes.

Let's return to our original example involving Rita, Charles, and Leah and pick up where we left off. Remember the squirrel, horse, and goat had returned to their sandpile home to be safe during the storm.

"I'm going to the big part," says Charles as he takes his squirrel from the sand, places it on top of the pile and begins covering it again. Leah and Rita help him.

Charles now pretends to pick up something from the sand (he cups his hand to suggest he has something in it) and places it at the far end of the sandbox, away from the house. He then shouts, "Hey creature! Don't go in the house. That's a snake. That's the snake—that wanted to go into the house."

Leah now takes a toy cow from the sand pile and moves it up and down, yelling, "Hey! I'm cold. Cold. I'm cold."

"Get into this pile!" commands Charles as he covers Leah's cow.

"Yeah," says Rita as she helps to cover the cow with sand.

"No! Don't get it off the top—sand off—or else our house will break down," advises Charles.

"Yeah—yeah, yeah! Get more," says Rita. "The faster we get, the faster we can get the sand away!" Rita is helping Charles and Leah, who are raking sand from all around the box to reinforce the house.

"Yes," says Charles. "The faster we push—the—the snow over, the faster we'll get the warm!" (The structure of Charles's turn is very similar to Rita's and it is said in the same cadence.)

"The—the sun goes on," announces Rita. "Whoopee! Whoopee!"

"Hey! The rainstorm is over!" shouts Charles.

"Yea! Whoopee! Get out!" screeches Rita. She takes her horse from the sand and holds it high in the air.

"Out," says Leah, as she and Charles reach in the sandpile and take out their animals.

The danger-rescue theme in this and all other instances I observed contained three phases. Each phase displays a different feature of peer culture regarding children's perceptions of danger. The first phase entails the recognition of danger. What is most interesting is how the danger evolves. Although kids expect danger to occur in spontaneous fantasy, its arrival is always a surprise. One must be on the lookout! Danger can come from anywhere and out of nowhere.

In this example the kids first build a home to escape the cold. Then it starts to rain. The rain becomes a storm. The storm includes lightning, and now there is a need for help. After the storm a snake tried to get into the house. Note that the children take no risks here. The danger that arises is not the result of reckless behavior. Rather, danger *is something that happens to children.* In peer culture, kids share a concern about danger, and they see it as something that can occur at any time.

Because danger often occurs without warning in spontaneous fantasy, the children must be prepared to deal with it when it arrives. Their main strategy is not confrontation but *evasion*, and the second phase in the danger-rescue theme is to "avert the danger."

Once the danger arrives in our example, Charles immediately takes evasive action. He moves his squirrel away from the house because "lightning only hits a biggest—bigger—will hit our house cause it's the biggest thing." But Charles quickly rethinks his evacuation plan. He decides that the house is safe after all because it is made of steel and the lightning will just fall to the ground. The two girls agree and the children put their animals back in the house so that the lightning "won't get them."

Note that averting danger is something the children do together. It involves communication and cooperation. In averting danger, one must be calm and careful, and not take unnecessary risks. Thus, danger-rescue is somewhat of a misnomer. The children do not rescue one another; rather, they *collectively escape the danger*.

The third phase involves the recognition that the *danger has dissipated or gone away*. Danger often departs as quickly as it arrives. And the dissipation of danger, like its arrival, is something that happens to the kids. Danger comes, the kids avert it, and it disappears. The recognition of danger's dissipation brings about a *shared display of relief and joy*.

In our example, once the animals are safely inside the house, several limited lines of action that embellish the nature of the danger are played out. A snake, a second possible source of danger, is removed from the house. Although we might question why Charles introduced the snake, its presence is not that unusual. The pretend snake (there was no toy snake to animate), like the other animals, certainly could have wanted to enter the house to escape the storm. But because the snake was itself a threat, it was removed. During the storm, the children venture outside briefly to reinforce the house. It is during this stabilization of the shelter that Rita recognizes the dissipation of the danger—the storm is ending because the "sun goes on." Charles quickly takes up on Rita's recognition and announces that "the rainstorm is over." Now all of the kids share in the celebration of danger's departure by removing their animals from the house with shouts and cheers.

Lost-Found Theme

I have observed two types of lost-found themes. One involves the (purposeful or accidental) loss of a play object, which is followed by the search for, and discovery of, the lost object.

Joseph, Roger, and Denny (all about three and a half years old) are playing in the outside sandpile of the Berkeley preschool. When they start their play, they bury a toy boat deep in the sand. They then build a pile of sand on top of where they hid the boat and take turns jumping up and down on it. Suddenly the children fall to their knees and Denny says, "Now let's pat it, OK?"

"OK," says Roger and he and Denny pat the top of the sandpile.

"Pat it on the top." says Roger. Joseph now joins in the play.

Roger motions for Denny and Joseph, "Wait! I know how to pat it." He picks up a shovel and scoops sand off the top.

"Let's dig it again," says Denny.

"OK," replies Roger and he puts down the shovel.

"Let's dig it like this so we can have a cake," suggests Denny. The three boys are now digging with their hands in the sand.

"Yeah, cake!" agrees Joseph.

"A cake—," Roger starts to say.

Denny now discovers the boat and interrupts Roger, "We can see a boat—a boat!"

"A boat! This is our treasure!" shouts Joseph.

"Our treasure! Our treasure!" Denny repeats as the boys pull the boat from the sand. Later they bury and discover the boat three more times, marking their discovery with shouts of joy each time.

The kids display *genuine excitement and joy* in finding their treasure, which makes this activity important in the peer culture. Although hiding and discovering a toy boat might not arouse much excitement among older children, it is important to note that these preschool children have recently moved from Piaget's "sensory motor" to his "preoperational stage" of cognitive development. Piaget argues that in the sensory motor stage of the first two years of life or so, children do not realize that objects continue to exist in the same physical form when they are out of touch or sight. For example, a young infant does not pursue a ball that rolls under a chair because she does not realize that it still exists under there. In Piaget's preoperational stage children acquire object constancy—the knowledge that objects keep their same physical characteristics when out of our immediate senses. Objects do not just disappear and reappear magically. Object constancy is a recent acquisition for these children. Repeated enactments of play involving hiding and discovering objects have both magical and self-autonomous (that is, their realization that they have new skills) features for children. The kids are excited about their *mastery of the still somewhat magical behavioral routine in which objects are made to disappear and reappear.*

A second version of the lost-found theme involves the kids' creation of fantasy events in which characters they invent and animate *personally become lost.* Like the lost-object variant, there is a great deal of excitement and joy in being found. However, the initial period when the animated character becomes lost is much more intense and anxiety provoking. Let's look at another example of spontaneous play in the Berkeley preschool of the three children we discussed earlier, Rita, Charles, and Leah.

Rita has three horses and moves them to the end of the sandbox where Leah and Charles have placed their animals in the sandpile (home). Rita moves one of her horses up and down and calls out, "Help! Help! I'm off in the forest!"

"Come in here," Charles advises.

"In here," says Leah.

"Come in here! Come in here!" repeats Charles.

"I can't," yells Rita. "I'm lost."

"OK," says Charles as he reaches over, takes Rita's horse from her hand, and puts it in the sandpile.

"My friends, they'll get burnt," says Rita who is still animating the horse in the sandpile.

"I'm cold! Freezing!" yells Rita as she now animates a second horse and moves it toward the sandpile.

"Stay in here," says Charles as he takes the horse and puts it in the sandpile.

Rita now picks up the third horse and shrieks, "I'm freezing too! I'm freezing too!"

"Get in here!" commands Charles as he takes the third horse from Rita and places it in the sandpile.

"Get in here," says Charles as he now pats the top of the sandpile with all the animals inside.

"Warm!" shouts Rita.

The children's enactment of the personal lost-found theme generates interpersonal cooperation and support. Thus, spontaneous fantasy promotes the development of language and social skills plus a shared sense of trust among peers.

Another, more abstract, aspect of the personal lost-found theme is that it is a manifestation of an attempt to cope with an underlying fear of being lost and alone. Many preschool children have directly experienced, even if only briefly, the amorphous and almost overwhelming anxiety that results from being lost. If they have not experienced this anxiety firsthand, most preschool children have been warned of the danger by parents or have shared the experience vicariously through media (fairy tales and films). Enacting lost-found themes shares many features with the production of approach-avoidance play that we discussed earlier. In both aspects of the peer culture, children are able *to share and feel in control of* various dangers, fears, or threats to their safety.

Death-Rebirth Themes

Death-rebirth themes were composed of four phases: (1) announcement(s) of dying and death; (2) reaction to or certification of the announcement; (3) strategies to overcome the death; and (4) rebirth and celebration.

In some cases a child's announcement of death was ignored or disputed by peers, while in others it was taken up immediately and a sequence of death and rebirth was acted out. Let's consider an example of each type, again from the spontaneous fantasy play of the Berkeley children, Rita, Charles, and Leah.

The children have created a sequence in which a tidal wave comes and destroys their home. Charles pretends that his animals are floating in the water and says, "Now they're sailing away."

Rita lays her animals on their sides and announces, "We're dead! Help, we're dead!"

Charles and Leah ignore Rita, and Charles says, "Water—."

Rita cuts him off and repeats, "We're dead! We're dead! Help!"

"Water flattened it," says Charles referring to the animals' home. "They have to go to—."

Rita again cuts him off, "We're dead. We're dead! Help!"

"We're gonna send them under water cause they're sailing away in a cave," says Charles.

"My sheep are safe. My sheep are safe," says Leah.

Rita now pushes her animals, still on their sides, and again shouts, "We're all dead. Help!"

"I can't—," begins Charles, who then corrects himself and says, "You can't talk if they're dead."

"Oh well, Leah's talked when they was dead," counters Rita. "So mine have to talk when I'm dead." She then whispers, "I'm dead. Help. Help. I'm dead. Help."

"Here's our chimney! Here's the chimney," says Charles. He places some animals at the very top of the sandpile pretending the chimney of the home is sticking out above the water resulting from the flood.

In this sequence Charles and Leah ignore Rita's announcement that her animals are dead (apparently drowned in the flood from the tidal wave). Rita persists, however, and Charles eventually reacts negatively, noting that the animals Rita is animating "can't talk if they are dead." Rita responds that Leah's animals talked when *they* were dead, but I did not see this happen in previous play. In any case, Charles and Leah continue to ignore Rita and the death-rebirth theme fails to develop. It is interesting that Rita is unsuccessful even when she whispers rather than shouts her announcement after Charles's rejection.

Although it might appear that Charles's negative reaction to Rita's announcement is arbitrary, this is not necessarily the case. It is true that one could never announce one's own death if a rule of "no talking when you're dead" were strictly enforced in the play. However, in my observations of other instances of fantasy play, it seems to me that Rita's error was not her announcement of death, but rather her additional and repeated calls for help. These additions prompted a reaction, but in the process violated the original claim that the animals were, in fact, dead.

The following sequence, which is a continuation of the spontaneous fantasy play of Rita, Charles, and Leah, supports this interpretation and also illustrates a full enactment of the death-rebirth theme.

Charles lays his animal on the bottom of the sandbox near the sandpile and says, "Rabbit's dead."

Leah now lays her animal next to Charles's and also says, "Rabbit's dead."

"No," protests Charles, "only my rabbit's dead." He then picks up Leah's rabbit and gives it back to her.

"What's a matter?" asks Leah as she moves her rabbit and stands him up next to Charles's rabbit, which is still lying down.

Charles picks up his rabbit and stands it next to Leah's.

Rita takes her horse from the sandpile and lays it down on its side at the other end of the sandbox. She then announces, "Oh, my horse is dead. My horse is dead."

Charles hops his rabbit over near Rita's horse, "Hop! Hop! Hop!" he exclaims. He then bangs his rabbit on the ground next to Rita's horse. "If I bang on it. Bang! Bang! Bang! He'll be alive."

"Bang! Bang! Bang!" says Leah, who has also brought her rabbit over and bangs it on the ground.

"If I bang on it," repeats Charles, "it will be alive." He then hits his rabbit on the ground next to Rita's horse: "Bang! Bang! Bang!"

"You better—you bang on that bell on'em," says Rita, pointing to a microphone I have hung above the sandbox area. "He'll be alive and he'll open up."

"Bang on what?" asks Charles, confused.

Rita again points to the microphone and Charles and Leah turn to look. "Bang on that bell. If you—if you wake 'em up. Bang on that pretend bell."

"What bell?" asks Charles, still confused.

Rita now stands and points to the microphone. "That pretend bell. That—bell. That microphone."

"I don't see the bell," says Charles, who is still looking for it.

"That microphone," responds Rita, who is now pointing directly at it.

"Oh," says Charles, seeing the microphone.

"Not that," says Rita, changing her mind. "This. This horse." Rita picks up a larger horse and holds it while the smaller (dead) horse still lies on the ground.

Charles takes the horse and hits the other one saying, "Brrring! Brring! B—rring!"

Rita now moves both horses up and down. "Jump horse. Whoa! Jump. Whoa jump!" she shouts happily and repeats it three more times.

This sequence begins with Charles's announcement that his rabbit is dead. Leah copies his behavior. Charles reacts negatively to Leah, implying that only his rabbit can be dead because he announced it first. Leah does not seem to understand, and at this point Rita says that her horse is dead.

Although I cannot definitely infer the children's intentions, it seems that Charles decides to "give up" on his animal being dead and move on with the play by responding to Rita's announcement. It is interesting here that unlike her earlier action, Rita only announces the death and does not call for help. In any case, Charles confirms the death and offers a strategy for bringing Rita's horse back to life (the third phase of the death-rebirth theme). In this instance, the third phase is fairly lengthy because there is debate about how to bring the horse back to life. Charles suggests banging on the ground next to Rita, but Rita wants Charles to bang on her horse with a bell. She points to a microphone that could serve as a bell. Because the microphone is clearly out of reach, Rita notes that it can be a "pretend bell." It is worth noting that Rita is talking even though she is dead. However, here the fantasy frame is temporarily broken as Rita talks for herself, using stage direction and the word "pretend" to accomplish agreement on how to bring the horse back to life.

The problem is solved when Rita gives up on pretending the microphone is a bell and offers a large horse to serve the purpose. Charles obliges and Rita's horse is reborn. She marks this with a great deal of joy through words and actions by having her horse jump around so happily that he has to be held in check by her saying "Whoa! Whoa!"

Although the kids' play demonstrates that they have knowledge of and talk about death, it is difficult to infer the degree of their concerns and anxieties. Surely, children think about death, and they are frequently exposed to information about illness, dying, and death by the media (especially television, movies, and fairy tales). In fact, this example of how the horse is brought back to life is similar to death-rebirth themes in fairy tales and Disney movies like *Sleeping Beauty* and *Snow White*.

The kids' production of death-rebirth themes in spontaneous fantasy enables them to share concerns or fears they have about death. Therefore, the theme is similar to personal lost-found and danger-rescue themes in spontaneous fantasy and also to approach-avoidance play. However, there is less tension in the death-rebirth theme. What the kids stress are the tactics involved in bringing the dead back to life. In this sense, the death-rebirth theme has an enchanting quality that children like: first enacting the theme and then sharing in the joy of the magical outcome.

THE ALMOST-PUPPET SHOW AND
THE ROCK 'N' ROLL SLIDE

In the preschools I observed, most spontaneous fantasy occurred in sand play or play with small building materials like Legos. Like the examples given earlier, the play involved the children animating the materials and other toy objects (animals, cars, spaceships, and so on) and producing fantasy events. However, in some schools children played with larger building materials like wooden blocks and planks. The children often used these materials to build houses, spaceships, hideouts, and so on, and in the process they spontaneously created fantasy play themes. This play differs from animating toy objects, in that the children themselves take on or embody certain characters and perform a variety of coordinated activities. Some themes in such activities were similar to dramatic role-play events, which I discuss in the next chapter. However, the children often produced events that had elements of real-life roles but also fantasy actions and themes that the kids made up on the spot and extended and embellished in the play.

Once in the Berkeley preschool a boy, Daniel, had the idea of producing a puppet show. He coaxed me and several children into being the audience, seating us on a carpet in front of a small bookcase. He gave each of us two small wooden blocks (one a bar of candy, the other a flashlight we were to use to find our imaginary seats).

"Now sit down and get ready for the show!" Daniel told us.

Then he and another boy, Tommy, went behind the bookcase and began banging away with hammers, ostensibly building the set for the show. After about 10 minutes of banging, Daniel reappeared and said that the show was about to begin.

Then there was more banging and the audience began to dwindle. Meanwhile, I sat and waited patiently with Sue, Sheila, and Christopher. Finally, Daniel and Tommy pushed two chairs up behind the bookcase, climbed onto them, and called for our attention. At last, the show was about to begin.

But instead, Daniel announced, "Tommy messed everything up" by not letting him use the good hammer, and that "the puppet show is canceled." Tommy denied this accusation and began pushing Daniel, and the two fell from their chairs. Tommy hurt his leg and began to cry, and a teacher ran over to help. I remained seated, but the other children got up to leave, with Sheila dropping her candy and flashlight to the floor and declaring: "What a gyp!"

In another example, in a private, upper-middle-class preschool in Bloomington, Indiana, several kids, all about five years old, took blocks and planks from the storage area and began to make a structure in the adjoining play area. At first, the kids positioned several blocks in a squared pattern. Then one boy, Doug, placed a plank across the top of two of the blocks. Without verbal negotiation several other boys, Andy, Scott, Larry, Bill, and Mark, also placed planks in the same pattern as Doug.

Before long all the blocks were covered with planks, and Doug yelled, "Hey guys, we made a stage!"

"Yeah," responded Larry and then he, Mark, and Bill got up on the stage and started dancing around.

Scott now came over carrying a wooden plank. He jumped up on the stage and danced around, strumming the plank as if it were a guitar. "Hey," he shouted, "We can have a rock 'n' roll show!"

"Yeah," said Doug, "but first we have to make the seats."

The boys then brought several blocks over and set them in front of the stage. I took a seat and waited for the show to begin.

"Now we need curtains and microphones," said Larry.

"OK, here's the curtains," agreed Bill as he held up his hands and spread them out, as if closing curtains around the stage.

Scott and Andy now pretended to plug in microphone wires on the stage and then Doug, Scott, Larry, and Andy stood behind the pretend microphones and played their guitars, making electric guitar-like sounds. Meanwhile, Bill opened the pretend curtains and added some more blocks to the audience area, but I was still the only one watching.

"Quiet," yelled Doug. The other boys stopped playing and Doug shouted out, "Welcome to our show!"

I clapped and the boys launched into another musical number. Scott ran up and down the stage with a Chuck Berry-type strut. I assumed he had seen this in a movie or on television.

While the band was playing, Bill piled three blocks at the far back of the stage and angled a plank from the top of the blocks to the stage floor. He then slid down the plank.

"Hey," said Doug, "it's the rock 'n' roll slide!"

All the boys dropped their instruments and took a turn going down the slide. Often they jumped up after hitting the stage floor, grabbed their guitars, played a few riffs, dropped the guitars, and headed back to the slide.

As they went down the slide they yelled, "Whee! Rock 'n' roll!" and "Rock 'n' roll slide."

I was pretty impressed with all of this and jotted down a lot of descriptive notes in my notebook.

Pretty soon the boys began doing what they called "cool stunts," like sliding down backward with their eyes closed and so on. Then Larry took his guitar (plank) to the slide, sat on it, and slid down. Several of the other boys copied Larry's stunt.

Later, Mark placed a plank with a small block at the base of the slide on the stage. Larry slid down and his foot hit the small block, launching it into the air. The boys were very happy with this stunt and repeated it several times. I felt it was a little dangerous and was relieved when one of the teachers came over and suggested that they use a stuffed rabbit instead of a block. The boys readily agreed.

However, soon after that they replaced the rabbit with another block. They then set rules that you were not to kick the block too far and that the one who tipped over the slide (this happened quite often) had to repair it for the next kid. These rules worked for a while and then Doug and Mark started kicking the block a long distance again.

"You can cheat in this game," said Mark.

Andy disagreed and threatened to tell the teacher. But Mark countered, "This is our game, so we can cheat if we want."

During the boys' play, a girl, Mary, came up on the stage and put her feet on two loose planks and pretended to ski. The boys laughed and some imitated Mary, who soon left the area.

At one point, Scott hurt his finger going down the slide and he went over to show the injury to Bill. Bill had been moving the boxes (chairs) around and playing some guitar but had not gone down the slide.

"I bet that really hurt," Bill comforted Scott.

Scott sniffed a bit but kept from crying and toughed it out. He even returned to go down the "rock 'n' roll" slide a few more times.

Finally, the teachers flashed the lights off and on, signaling "clean-up time." As the kids returned the blocks and planks to their places in the storage area, I picked up a plank. I did a little Eric Clapton air guitar imitation, but nobody noticed. I really wanted to go down the "rock 'n' roll" slide, but I was too big.

5

"When I Grow Up and You Grow Up, We'll Be the Bosses"

. . . .

Role-Play in Kids' Culture

Two five-year-old girls, Jean and Karen, are pretending to have a tea party in Jean's home. Jean's mother has told them that they can each have one of three types of cookies during their play. The girls have finished their three cookies each, but two more cookies remain on the plate.

"Let's pretend," says Jean, "When Mommy's out 'till later, OK? And these two can be off and she didn't want we eat one—and we pretend we ate it later, OK?"

"Oooh," whispers Karen. "Well, I'm not the boss around here though. 'Cause Mommies play the bosses around here."

"Yeah," says Jean.

"And us children aren't," declares Karen, shaking her head.

"So long ago," says Jean. She seems to lose her train of thought, but then adds, "When I grow up and you grow up, we'll be the bosses!"

In socio-dramatic role-play of the type produced by Jean and Karen, kids collaboratively produce pretend activities that are related to experiences from their real lives (for example, family and occupational roles and routine activities) as distinct from fantasy play based

111

on fictional narratives like we saw in Chapter 4. Child researchers have long argued the importance of dramatic role-play for children's social and emotional development. Like most adults, these researchers most often see role-play as the direct imitation of adult models. However, kids do not simply imitate adult models in their role-play; rather they continually elaborate and embellish adult models to address their own concerns.

Kids' appropriation and embellishment of adult models is primarily about status, power, and control. Kids are empowered when they take on adult roles. They use the dramatic license of imaginative play to project to the future—a time when they will be in charge and in control of themselves and others.

Role-play also allows kids to experiment with how different types of people in society act and how they relate to each other. Of great importance here for children are gender and expectations about how girls and boys should act and how roles in society are gender stereotyped. We will see that young children do not accept but challenge and refine such stereotypes. Thus, gender role expectations are not simply inculcated into children by adults; rather, they are socially constructed by children in their interactions with adults and each other.

"GET DOWN IN THE BACKYARD YOU TWO CATS": ROLE-PLAY AND SOCIAL POWER

Kids begin role-play as young as age two and most role-play among two- to five-year-olds is about the expression of power. In my dissertation research I was interested in language used in the play of a brother and sister, Krister and Mia, and a second boy, Buddy. In one play session, Mia (who was four and had been to preschool) and the two boys (both about two and a half years old and without preschool experience) began a role-play sequence when Mia suggested that we play teacher. Krister wanted to be the teacher and pushed a chair to the front of a large blackboard in the room. Mia, Buddy, and I sat on the floor as students.

Krister took the chalk and said, "Now write this!" and drew several lines.

"Those aren't letters, but just a bunch of lines!" I responded teasingly.

"He can't write so good," Mia told me, a bit annoyed. "Just pretend they're letters."

But Krister did not allow his authority to be tested. He shouted out at me, "Bill, you are bad! You must go sit in the corner right now!" Krister pointed to the corner of the room, and I took my paper and went over there to sit. Buddy and Mia began to laugh, but Krister gave some more orders about what to write and Mia, Buddy, and I did what we were told.

Here we see a young child who had not attended preschool but had information that teachers are powerful and tell kids what to do. Also, bad kids are made to sit in the corner. Did Krister learn this from Mia? Possibly, but not as a result of her own experiences in preschool. Their father assured me there was no sitting in the corner in Mia's school. Perhaps it was from something on television such as a cartoon or an adult joking about kids having to sit in the corner if they are bad in school. Where Krister picked up the information is less important than his desire to express the power one has in an adult or superordinate role (that is, a role with the most power or authority), a situation in which young children seldom find themselves.

In socio-dramatic play, children relish taking on and expressing power. It's fun. In one complex role-play episode from my work in Berkeley the kids (all about four years old) clearly expressed power and control while in superordinate roles, misbehaved and obeyed in subordinate roles, cooperated in roles of equal status, but became confused about the alignment and gender expectations of other roles.

A boy, Bill, and a girl, Rita, entered the upstairs playhouse carrying purses and a suitcase. Before coming upstairs they had agreed on the roles of husband and wife. As they dropped the purses and suitcase on the floor, they looked down at children playing below. They saw two boys, Charles and Denny, crawling around and meowing like cats.

"Hey, there are our kitties," said Bill.

Rita replied, "Yeah, they're down in the backyard."

What is interesting about this simple exchange is how much it accomplishes. Before this talk, there had been no discussion of kitties, nor had Rita and Bill talked to the two boys. However, Bill and Rita might have presumed that the boys would come upstairs and expecting this, they *made* them their two kitties and made the downstairs the backyard merely by saying it was so. Here we see that much role-play, like the spontaneous fantasy play that we saw in Chapter 4, is improvised.

Bill and Rita now went about arranging things in the house. They picked up blankets from the bed and placed the purses and suitcase on the floor in front of the bed. Bill then picked up a baby crib and placed it alongside the front of the bed, blocking off the area around the bed from the rest of the room.

"This is our special room, right?" said Bill.

"Right," responded Rita.

"This is our little room we sleep in, right?" added Bill. "Our little room. Our—."

"We're the kitty family," said Denny cutting off Bill as he and Charles climbed up the stairs and into the playhouse. They began crawling around the room, meowing.

"Here kitty-kitty, here kitty-kitty," said Rita, reaching out to pet them. "Yeah, here's our two kitties," she announced to Bill.

"Kitty, you can't come into this room!" Bill commanded sternly. But one of the kitties, Charles, immediately crawled into the room and climbed on the bed. Meanwhile, the other kitty knocked a plate from the table to the floor.

"No! No!" yelled Bill. He then shooed the kitties back toward the stairs. "Go on! Get down in the backyard!"

Rita came to Bill's aid and shouted, "Get down in the backyard, you two cats! Go down! Down! Down!"

The kitties headed toward the stairs and Charles started crawling down. But Denny stopped at the head of the stairs and said, "No, I'm the kitty. I'm the kitty." It seemed that he wanted to stay. But the husband and wife insisted that he go.

"Go back in the backyard!" commanded Bill.

"You get in the backyard. Ya! Ya!" yelled Rita, pushing at the remaining kitty with her hands.

Denny now gave up and also went down the stairs.

Bill looked down at the two cats and said, "Go in the backyard. We're busy!"

"They were rough on us," said Rita.

In this sequence we see that the husband and wife express clear authority over the kitties through their use of imperatives expressed with strong intonation and accompanying gestures of control. But we also see that the kitties brought on these strong displays by their misbehavior and resistance. In fact, in many role-play episodes, subordinates (kids or pets) often misbehaved by doing exactly what they were told not to do! In the process, discipline scripts emerge with a language structure like we just saw, in which power is clearly displayed and enforced. It is as if the kids want this to happen. They want to create and share emotionally in the power and control adults have over them.

After the kitties left, the husband and wife decided that the house needed cleaning. In line with stereotyped gender roles, Bill moved the furniture while his wife, Rita, cleaned the floor.

Bill picked up the table and said, "Be careful. I'm gonna move our table."

"You're a handyman, handyman," sang Rita.

"Next," said Bill as he pushed the stove near the door and then moved the table next to it.

"Bill? Bill?" called Rita.

"What?"

"You're a strong man," Rita praised him.

"I know it. I just moved this," said Bill referring to the table.

Here the children work together in line with stereotyped gender role expectations that are expressed in actions (that is, husbands are strong and help around the house to move furniture while wives do the cleaning) and reinforced in verbal evaluations (for example, Rita noting that Bill is a handy and strong man).

As Rita is pretending to mop the floor the kitties returned. Bill tried to block them off, but they scurried by, moving onto the just cleaned floor. Bill attempted to shoo the kitties back to the stairs.

"Come on kitties, get out! Get out! Scat! Scat."

Rita stopped cleaning to help her husband. "Come, scat. Scat!" she yelled.

Charles crawled back down the stairs, but Denny remained and stood up announcing, "I'm not—I'm not a kitty anymore."

"You're a husband?" Bill asked.

"Yeah," agreed Denny.

"Good. We need two husbands," said Bill.

Now Bill called out to Rita, who did not seem to hear the previous exchange. "Hey, two husbands."

Rita was not pleased with this development and offered an alternative. "I can't catch two husbands cause I have a grandma."

"Well, I—then I'm the husband," said Denny.

"Yeah, husbands! Husbands!" chanted Denny and Bill as they danced around the room.

"Hold it Bill," said Rita. "I can't have two husbands."

Rita held up two fingers and shook her head. "Not two. Not two." She then walked down the stairs. Meanwhile Bill and Denny continued dancing around upstairs and chanting "Two husbands! Two husbands!"

Rita walked around in front of the downstairs playhouse shaking her head. She stopped near the stairs just as Bill and Denny came down, and said, "I can't marry 'em, two husbands. I can't marry two husbands because I love them."

Bill said to Rita, "Yeah, we do." He then turned to Denny and said, "We gonna marry ourselves, right?"

"Right," responded Denny.

The boys then went back upstairs and continued chanting, "Husbands!" They danced around and jumped on the bed, but there was no coordinated activity. It was not clear to them or to me what two husbands do, especially without a wife. Later Rita came upstairs and said she was a kitty. The two husbands admonished her for scratching them and misbehaving and chased her down the stairs. Shortly after, the role-play was brought to an end with the teacher's announcement of "clean-up time."

In this sequence the role-play hit a snag, at least for Rita, when Denny decided he didn't want to be a kitty anymore. Perhaps he was getting tired of being shooed down the stairs. In any case, Bill suggested that Denny also be a husband and when Denny accepted, Bill even said, "Good. We need two husbands." It is not clear why Bill made this offer. Mostly likely because Denny is a boy and males are husbands, Bill thought that Denny should be a husband like him.

Rita, however, thought otherwise and saw a problem that goes beyond gender stereotypes: one wife and two husbands. While the boys danced around and celebrated being two husbands, Rita argued to no avail that she cannot catch, have, marry, or love two husbands. She knew that something was wrong with this relationship (at least among the adults in her culture). What was wrong has to do with her emerging knowledge that the roles of husband and wife are not only gender specific but also related to each other in particular ways. Wives and husbands love each other and get married. It is even assumed that is the case in her pretend relationship with Bill. But what was she to do with Denny?

She seemed to offer up the role of grandma for Denny. "I can't catch two husbands cause I have a grandma." But her phrasing is confusing and a grandma is the wrong gender; grandpa might have worked. The contrast of the boys' glee at being two husbands—Bill even suggested that they marry themselves but no such ceremony occurred—and Rita's discomfort with the proposed arrangement is interesting. In the end, she solved the problem by becoming a kitty and the play continued with a reversion back to misbehavior and discipline. However, Rita had a glimpse into the complexity of role relationships. In Piaget's terms, she had a disequilibrium in her sense of her social world, which she will strive to compensate for. So we see that role-play is fun, improvised, unpredictable, and ripe with opportunities for reflection and learning.

"*NON C'È ZUPPA INGLESE*": PLYING THE FRAME IN ROLE-PLAY

As I suggested above, role-play involves more than learning specific social knowledge; it also involves learning about the relationship between *context* and *behavior*. As the anthropologist Gregory Bateson argues, when the child plays a role, she or he not only learns something about that role's specific social position but "also learns that there is such a thing as a role." According to Bateson, the child "acquires a new view, partly flexible and partly rigid" and learns "the fact of stylistic flexibility and the fact that choice of style or role is related to the frame or context of behavior."

Children's recognition of the "transformative power" of play is an important element of peer culture. It is their use of this transformative power in role-play that I will, in line with Bateson and the sociologist

Erving Goffman, refer to as "plying the frame." Let's consider some examples.

In Bologna a girl, Emilia, has made an ice cream shop with two of her friends. She comes to where I am playing with three boys, Alberto, Alessio, and Stefano. I have a microphone in my hand because we are videotaping the play.

"Bill, will you come to see our store?" she asks.

"I can't now because—ah—I'm here with this—." I struggle with my answer not sure how to say what I need to in Italian.

"*Microfono*," she finishes my reply.

"Yes. I can't ah—," I say, motioning that the microphone wire is not long enough to go to her store. "Will you bring the ice cream to me?" I try to say, but my grammar is incorrect and she does not understand.

"What?"

"Take the ice—," I blurt out, confusing the words for "bring" and "take." But then I recover quickly. "Bring me the ice cream, to me."

"Yes. But we still have to—," she begins.

"Chocolate and—ah—chocolate and va—vanilla," I say. I noticed earlier that Emila and her friends were using dirt as pretend chocolate ice cream and sand for "*crema*" or vanilla.

"Yes," she says, "but we must finish the store. We still have to make it—the vanilla."

"Yes, that's fine."

"After I give it to you," she continues, "there's also strawberry. There is—I'll tell you all the flavors."

"Yes," I say.

Emilia gestures, counting off each flavor first with her thumb and then with fingers of her hand. "Eh, strawberry, chocolate, vanilla."

"Lemon?" asks Stefano.

"No, there is none," Emilia tells him.

I say, "I like ah—vanilla and ah—strawberry."

"Okay."

"For Stefano," I say, "for Stefano vanilla."

But Stefano wants to make his own order. "For me strawberry and banana."

Having just listed the flavors, Emilia is frustrated with this order. "There is no banana!" she insists. After all, this is a small ice cream store without many flavors because the girls are trying to use things like dirt and sand to make chocolate and vanilla, and perhaps leaves for pistachio. I am not sure what they are using for strawberry.

"Lemon," says Stefano, knowing full well there is none.

"There is none!" replies Emilia.

"There is no lemon," I remind Stefano.

"Chocolate," Stefano finally agrees.

"Chocolate," repeats Emilia as she heads toward her store to fetch the ice cream.

However, now Alberto places an order: "Hey, hey, for me, *zuppa inglese*—whipped cream and pistachio!"

Alberto's request for "*zuppa inglese*," a rare flavor derived from the English dessert trifle, is so outlandish that Stefano, Emilia, Alessio, and I break into fits of laughter. After all Emilia just went through this business with Stefano and his request for lemon.

"*Zuppa inglese*," Stefano and I say, laughing.

"They don't have it," I tell Alberto.

Emilia returns and bends over Alberto and says: "*Non c'è zuppa inglese, non c'è pistacchio!*" ("There is no zuppa inglese! There is no pistachio!")

"Okay, then, I'll take banana," says Alberto.

Now there are howls of laughter.

"There is none!" Emilia says with a big grin.

"Okay, then, I'll take whatever there is. Chocolate," Alberto finally agrees.

"There's chocolate. There's vanilla, chocolate, strawberry, maybe pistachio."

"Orange soda?" asks Alberto.

"Well, I'll go see," says Emilia and she returns to her store.

In this example, Emilia at first wants to stay in the confined frame of pretending to have a small ice cream store with flavors that can be represented by features on the playground: dirt, sand, leaves, and so on. Although I have trouble making my order because of my fractured Italian, I stay within the frame and accept, no, even volunteer, "chocolate," a flavor I know she has. But first Stefano and then Alberto more or less say, "What's the fun of that?" They ply or stretch the frame by purposely ordering flavors that they know Emilia doesn't have or doesn't want to pretend to have. Then the whole role-play becomes about "playing with the play."

This turn of events is most apparent when Alberto calls out after Emilia as she is leaving and orders *zuppa inglese*. Now even I get what is going on and join in the laughter of the other boys at Alberto's request. Emilia, feigning disgust, clearly enjoys dealing with Alberto. She relishes the opportunity of denying the request, by responding "*Non c'è zuppa inglese!*" But Alberto's response to this is to ask for banana! Later, however, Emilia gives in some and says there might be some pistachio and she will check into the orange soda.

My Italian colleagues and friends loved to hear this story about the *zuppa inglese* and the inventiveness of the children's play. Once I told the story to friends at dinner while the waiter was passing by with the dessert cart. As I ended my story with, ". . . then Emilia said, '*Non c'è zuppa inglese!*'" to the amusement of my friends, the waiter responded, "*Ma signore, c'è zuppa inglese, c'è!*" pointing to a large bowl of trifle on the cart.

Another way children frequently ply the frame in role-play brings us back to the issue of gender expectations. As we saw in the earlier example of the two husbands, Rita, the original wife in the play, re-fused to stretch the play frame to allow herself to be the wife of two husbands. In the end she took the role of kitty and the husbands or-dered her around. However, this segment of the play was short lived, because the husbands seemed to have little idea what to do with each other besides dance around.

Although I did see girls sometimes take male roles in socio-dramatic play, boys almost always refused female roles. In one example from the Berkeley preschool, two five-year-old boys, Graham and Pe-ter, had entered the upstairs playhouse and were sitting at the table.

"You be the mommy and I'll be the daddy," said Graham.

"No, mommies are girls," Peter replied.

There was a long silence as the boys sat looking at each other.

"I know," said Graham, "I'll be the daddy and you be the uncle."

"OK," said Peter.

There was another long silence as the boys thought about what a daddy and an uncle would do. Finally Peter looked under the table and noticed that the round wood top was attached to the bottom of a metal waste can.

"Hey, look. It's a trash can."

"Let's flip it over," suggested Graham.

The boys flipped over the table and Graham got inside the can. Peter then pushed him around the playhouse, going "Vroom! Vroom!" So we have a daddy and an uncle out for an afternoon drive.

Several ethnographic studies of children's peer culture have docu-mented that older preschool children often extend and embellish tra-ditional socio-dramatic play. The psychologist Steven Kane, for ex-ample, found a decrease in traditional socio-dramatic play like family and occupational role-play in a group of four- to five-year-old children he studied over the course of a year. The traditional role-play was re-placed by what Kane termed "imaginative role-play" involving animal families or fantasy characters like royalty and knights.

The older children (five- and six-year-olds) I studied frequently produced animal family play in which they embodied wild and aggressive animals. In the play, the baby animals had much more freedom of mobility and were more aggressive than children in human family role-play. Further, animal mothers had fewer rules for their children but were more likely to mete out frequent physical punishment. In my work in Modena, Italy, the five- to six-year-old kids rarely engaged in traditional family role-play. Instead, they frequently produced a play routine in which a group of boys and girls pretended to be baby and adolescent wild dogs and lions that roamed around the school growling and scratching at each other and other children playing nearby. This pack was normally disciplined by one or two mother animals that were even more physical and aggressive than their charges. Let's consider a typical example from my field notes.

Several kids (Valerio, Angelo, Viviana, Carlotta, Daniele, and Luciano) are pretending to be wild dogs moving in and out of a corner of the classroom that they have closed off and use as their house or den. They growl and scratch at each other and other children not involved in the play. Federica is the mother dog and a strict disciplinarian. She swats at her children and yells, "*Adesso basta!*" (That's enough!). This warning is frequently used by the teachers and the children's parents when they are very serious.

Although some of the dogs (Luciano, Angelo, and Carlotta) are good and do what they are told and go to sleep, others misbehave. Daniele, for example, ignores Federica and even scratches her. Valerio and Viviana are also disobedient and constantly in trouble. Federica shouts at them loudly and swats them hard as they cower in a corner of the den. The play is so realistic that I am not sure that Federica, Valerio, and Viviana are not really upset.

Valerio and Viviana whimper loudly while Federica, who is now very red in the face, falls exhausted into a chair, exasperated with her unruly children. Again, I am having difficulty deciding if the play is getting out of hand, and if the children are really upset with one another.

"Are they mad?" I ask Sonia who is drawing at a table and not part of the role-play.

Sonia laughs at my question as if my concern attests to my having been fooled by her peers.

"The baby dogs are bad and Federica is a little severe," she tells me as she continues drawing.

I see that the teachers are aware of the loud play but do not comment, nor do any of the children who seem upset go to the teachers to complain or ask for help. When the teachers announce that it is time for lunch, Viviana still seems a bit upset to me. She accepts my hug, but smiles and says she's fine.

The embellishment of traditional family role-play in animal families leads to heightened aggression and emotion in the play. At one point the line between pretend and reality becomes blurred when some of the kids, at least as far as I could tell, become upset with one another. However, my concern about the children's brief distress was not shared by the teachers, their uninvolved peers, or the kids themselves.

For the kids, animal family play is not simply a set of scripts to be enacted, but a stretching or plying of the general family role-play frame. This plying of the frame gives the kids more control over it and enables them to move it in directions in line with values and concerns of the peer culture. For example, young animals have more freedom and can be more aggressive than human children. The mother animal is also more physically and verbally aggressive than human mothers. At the same time, however, such stretching of the frame can be unpredictable and risky: The pretend aggression and injury can become too close to the real thing. But it is this very risk that makes the embellishment of the traditional role-play so attractive to the kids. We see here an emotional element of role-play that is often overlooked in traditional cognitive developmental interpretations of such play. The children explore their own emotional reactions in their roles as parents and children. Also, both boys and girls engage in and test the limits of bodily and social aggression. Again we see that play is never simply play in children's peer cultures.

THE ICE CREAM STORE AND TELEPHONE TALK: SOCIAL REPRODUCTION IN ROLE-PLAY

The fact that children's role-play is often more than meets the eyes of adults is especially evident when we compare children's role-play across social-class groups. Some child researchers have argued that lower-class children are lacking in fantasy and role-play skills and need training to develop skills to engage in such play. Head Start, a compensatory program for economically disadvantaged preschool-age children in the United States, is based on such a notion of "deficiency" in these children's play and language skills. Thus, it is assumed they need a "head start" to catch up with middle- and upper-class children before elementary school. Other researchers have challenged this "deficit model" and argue that lower-class children's language and role-play skills might differ in various ways from those of the middle class but that they are not deficient.

In my own work with minority children in Head Start programs, I have found that the kids' role-play is highly creative in terms of language use and interactive and cognitive skills. It does, however, differ from the role-play of middle- and upper-class children I studied in that it is highly realistic and stays very close to details of the adult model.

So far, in discussing kids' role-play we have considered what it tells us about their developing knowledge of status, power, role alignments, and gender expectations. We have also seen how kids ply or stretch the role-play frame to embellish or "play with the play." In this way, kids are more in control of their play and use it to address concerns in peer culture and simply to have fun.

Role-play is fun for kids and while they are having fun they are also creating images of the adult world and reflecting on their place in that world in the present as well as projecting to their futures as adults. Therefore, in role-play, kids link or articulate local features of the ongoing play to their developing conceptions of the adult world. This articulation enables them to appropriate aspects of the adult culture, which they use, refine, and expand. It is through such appropriation that the children extend their peer cultures and contribute to reproduction of the adult world. This is a process I have referred to as "interpretive reproduction" (children *actively* contributing to the reproduction of adult society through their activities in their own peer cultures). A comparison of the role-play of a group of upper-middle-class children with that of economically disadvantaged children helps us capture this idea of interpretive reproduction.

In addition to the research sites I discussed in Chapter 1, I also observed children over long periods in two private, not-for-profit preschools and a Head Start program in Bloomington, Indiana. Let's compare role-play in one of the Bloomington private preschools, which I'll call University Preschool, with role-play among the Indianapolis Head Start children.

The kids at University Preschool frequently engaged in socio-dramatic play primarily by adopting family and occupational roles. In the following example, several children are "making things" as they stand around a sand table in the outside yard of the school. At one point, a child makes a reference to ice cream and then the four children (Ann, Linda, Tom, and Ruth, all about five years old) decide that they are owners of the ice cream store. I am sitting nearby with a microphone, because we are videotaping the play. We begin as Ruth enters the play.

"Hey, I heard you guys are making ice cream," says Ruth.

"We're making rainbow ice cream," Linda replies.

"Oh, rainbow ice cream," says Ruth.

"That's the best," I add.

"I'm making silly unicorn rainbow ice cream," says Ann.

In this sequence we see that after the kids arrive at an understanding that they are making ice cream, they quickly link the activity to their peer culture. They do this by the creative reference to making a flavor of ice cream that is similar to a toy (a rainbow unicorn) that many of the girls own.

"I know—I know what you can be," Linda proposes. "Um, I know—I—you—Ann—Ann, you and me and Tom and Ruth, we could all be the owners of this store and he could be the customer."

"I'm the customer," I agree.

"Bill's the customer," Linda confirms.

"Okay," I say, "I got a big order for you guys."

"What?" asks Linda.

"I want three chocolate ice cream cones, one quart of rainbow ice cream, and two pints of vanilla."

"Oh, that's a lot of work," says Linda. "You'll need to wait a long time for that."

"But I'm in a hurry," I protest.

"Now we only have chocolate ice cream," says Ann. "We don't have no—."

Linda now hands me a container filled with sand. "Here's the rainbow ice cream."

"Ok," I say and set it on the ledge of the sand table.

"With a cherry on top!" adds Linda.

Tom now says, "You have to be the ice cream—," but he is cut off by Ruth.

"It's gonna take us a long time."

"Yeah," Linda agrees, "'cause we can't make so much in one time."

"Okay," says Ann.

"Yeah," adds Ruth, "even though we all have ice cream. How about I put it in there and you take it out, alright?"

"No," says Ann, "let me get some chocolate ice cream. The, hmmm, let's see—."

"They are not melted," says Ruth referring to the ice cream orders.

"They're not melted?" I ask.

"Yeah," says Linda. "If you—this is a special kind of ice cream, that even if it stays in the sun for a long time it won't melt."

In this sequence, the children's class backgrounds are surely important in their defining of themselves as owners of the ice cream store as opposed to the more common alignment I have seen in such play, that of being workers and bosses. Once the definition of owners is accepted, the kids work together to fill my order. However, as is often the case in role-play, there is shifting back and forth from adult to peer culture. For example, after acting like co-owners and coordinating their work, the kids surmount the real-world problem of melting ice cream (it takes time to fill such a big order and it's hot outside) through the magic of pretending. Their ice cream is a special kind that won't melt "even if it stays in the sun a long time." Later in the play I ask how much money my order will cost.

"Um, three dollars," says Tom.

"Yeah, three dollars," agrees Ruth.

"Yeah, three dollars," echoes Linda.

"Who gets the money?" I ask.

"Me," says Linda.

I start to count out the money, "One—."

"Now remember this, remember this," says Linda, "remember that this goes to the hospital."

"It goes to the hospital?" I ask, a bit confused.

"Yeah," says Linda.

"The money does?" I ask.

"Yeah," Linda agrees.

"For charity?" I ask. I am still not sure what she is proposing.

"It's to help kids," says Linda.

"Help kids in the hospital?" I ask.

"Yeah," Linda responds.

Although Linda implies that the decision about giving the money to the hospital had been made earlier ("Now remember this—"), there has been no such discussion or reference to this proposal earlier in the play. In fact, Linda is improvising this line of action in the play through her skillful use of language. She asks us all to think back and remember an imaginary time when the owners agreed that money paid for the ice cream would go to the hospital to help sick kids.

There are several things to note here. First, as was the case in defining themselves as owners of the store, the decision to donate money for sick kids is surely also related to the kids' experiences in their families. Given that they are from upper-middle-class families, they have most likely attended or been exposed to the idea of charity events. Second, although Linda introduces the idea and talks with me about it, as we will see, both Ruth and Ann pick up on the proposal and expand on it later. Finally, we see differences in the kids' developing concepts of the world and their language as compared to an adult's. I use the word "charity" while the kids talk about "helping kids." Let's consider a final segment from the role-play.

"Here's your—another ice cream cone," Linda says as she hands me a plastic scoop filled with sand. She then informs me: "You can stay here for day and night, without stopping eating, 'cause we can work day and night."

"You guys work 24 hours a day?"

"Yep. No, we work 24 hours the day and night!" says Linda.

"We work all the time," says Ann.

"Yeah," adds Ruth, "we never stop working."

"You never have a break?" I ask.

"No. We don't want to," says Linda.

"We have to work all the time," Ann adds.

"All night and all day, 'cause we have to pay money for the hospital a lot, to help the kids," says Ruth.

"That's right," I say. "I forgot about that."

"Yeah, I said that this money goes to the hospital," Linda reminds me.

"Yeah, to help the kids," adds Ruth.

"Sick kids?" I ask.

"Yeah," replies Linda.

"Yeah," says Ruth, "but all—and all goes to the sick kids, 'cause if you look at the sick kids don't have very much money because the hospitals take it all away!"

"I'm making some drinks," says Ann.

"They have to use their money to pay the hospital bills?" I ask Ruth.

"Yeah," she answers. "So we send the money to the hospital to give to the sick kids. And sometimes we even send balloons for the kids that are being good."

The sequence begins with the kids talking about having to work a lot, "24 hours the day and night!" This discussion of long hours and hard work prompts Ruth to return to the idea of money for sick kids. She expands the idea further with the gist of her argument being that hard work and investment of time yields money (or profits) which is needed to help kids who are both sick and, therefore, also economically disadvantaged because of hospital bills. However, even in developing this highly sophisticated analysis, Ruth also retains important elements of the peer culture in that she notes that the sick kids will get not only money but, if they are good, also balloons.

Overall, we can see that in the security of their role-play the kids connect aspects of the adult world and their peer culture. In the process, they create windows through which the future is foreseen. In this way, the production of the routine itself contributes to the eventual production of aspects of the wider adult culture. We will return to this point after we consider an example of role-play among the Head Start kids.

As was the case in University Preschool, the kids in the Indianapolis Head Start center frequently produced socio-dramatic play that recreated family and occupational roles. Let's consider an example in which two girls (Zena and Debra) pretend to be mothers talking on the telephone in the family living area of the classroom. As we will see, the telephone talk involves general themes about the difficulties of parenting. The talk is impressive because the girls are producing their own interpretation of their mothers' telephone conversations about their (the mothers') parenting demands.

Telephone narratives of this type often involve not only the reconstruction of past events but also evaluations and interpretations of these events by both the tellers and the audience. In this way they both reflect and constitute shared culture. As Peggy Miller and Barbara Moore argue, when "caregivers habitually tell and retell personal stories, they are constantly reminding themselves of the experiences that are meaningful to them and relevant to their child-rearing beliefs and practices."

We can now turn to Zena and Debra's role-play. They have toy telephones and first pretend to be the women bus drivers conversing about the kids who ride on their buses—which ones are good and which cause trouble. They then say goodbye and hang up. Debra redials her phone and Zena answers.

"What you been doin'?" asks Debra.

"Hah. Cookin'. Now I need to go to the grocery store."

"I got to take my kids to the party store, they told me, I said—."

"My kids," interrupts Zena, "my kids want me to take them to the park!"

"What?"

"My kids told me to take them to the park," continues Zena, "and then, and then the bus had to come and get 'em. That's gonna be a long walk for to here! And then the bus would have to come and get us!"

"Well," answers Debra, "we have to wait for transfers, then I have to buy groceries. We have to buy some groceries. And um—."

"Guess where my kids told me to take them?" asks Zena excitedly.

"To the store. When the bus come by my kids waitin' for it. I don't got time to do that."

In this sequence the girls skillfully build coherent discourse through what the anthropologist Marjorie Goodwin terms "format tying" (the repetition of certain words or phrases of prior turns and semantic links across turns) regarding their pretend kids' requests, to construct the shared topic of problems of parenting in poverty. For example, in her answer to Debra's question of what she is doing, Zena notes she has to go to the grocery store. Debra builds on this syntactic element by noting that she has to take her kids to the party store. Zena then picks up on the talk about kids and says that her kids want her to take them to the park (what Goodwin means by semantic linking). In later turns the girls discuss the difficulties of doing these things and develop the theme of parenting in poverty.

The content of the talk as well as the structured order of turn-taking is also important. It is not just that the mothers (animated by the children) have to do everyday chores like shopping, their children also expect them to provide additional services. For example, the kids want to go to the party store. The party store is a type of small business that carries fewer items at higher cost than large grocery stores. In poor neighborhoods in inner cities, there are few grocery stores and residents try to keep their reliance on party stores to a minimum, that is, for basic necessities. However, this problem is a difficult one for young children to understand and Debra captures this difficulty in noting her kids *told* (rather than *asked*) her to take them there. Further, the girls' discourse captures their mothers' frustrations in trying to meet their children's demands to take them to the park and other places when they don't have a car and must deal with a limited and time-consuming bus service.

Later in the episode, the children continue to talk about the difficulty of parenting, noting numerous occasions of misbehavior of their pretend children. This misbehavior leads to reprimands and physical punishment, but the kids still misbehave. In fact, the girls pretend that their children are making so much noise at the moment of their telephone conversation that they have trouble hearing each other. At one point, Debra even covers the phone receiver to shout out to her pretend children to be quiet.

After the talk about discipline, Zena, who is standing some distance from a table where Debra and I are sitting, asks to talk to me. Debra hands me the phone.

"What are you talkin' about?" I ask Zena.

"Oh, we're talkin' about the kids, our kids are—."

"You got bad kids?" I ask.

"Very bad," says Zena. "I was gonna give 'em some ice cream but I can't. And I told them that I would."

"Told 'em what?"

"I told 'em, I told 'em, 'be quiet, be quiet.' But they wouldn't listen to me."

"And then they got some ice cream?" I ask.

"No!" shouts Zena.

Debra now speaks up without using the phone. "You shouldn't do that," she says. Here she means give ice cream to kids when they will not behave. Then she asks me, "Guess what my kids did? My kids said cuss words right in front of my momma!"

"Oh," I respond. "Who taught them those cuss words?"

"Probably cousins," says Zena.

"My sister's boyfriend," says Debra.

"That's where they heard the cuss words?" I ask.

Debra, frowning, nods her head.

Having overheard the earlier discussion about misbehavior, I ask Zena if she has bad kids. Zena says the kids were very bad and she could not give them promised ice cream because they would not obey and be quiet. Both she and Debra are emphatic about not giving in and letting the kids have ice cream as I suggest. Debra then relates an instance when her pretend kids were not only bad, but put her in a very embarrassing position by cussing in front of her mother. In response to my question of who taught her kids the cuss words, Debra exhibits intricate knowledge of her complex family structure and how it influences family interactions and parenting.

A final segment from the role-play reinforces the complex nature of these girls' family lives and their keen awareness of the stark realities of growing up and parenting in poverty.

Zena is again talking on the phone to Debra and is now seated at the table with her. She says, "You know what girl? My daughter asked me for pop. Every hour and all day. I say, 'No pop. You're gonna eat ice cream and cake and water—drink water and brush your teeth. Eat gum—.'"

"Guess what?" says Debra. "I'm getting ready to drive over to your house."

"I won't let you in," Zena responds.

Surprised by Zena's refusal, I ask "Why not?"

"'Cause," says Zena. But she changes her mind and tells Debra, "I'll let you in."

"My man start in on me," says Debra. "He's been hittin' on me. He's been hittin' on me for 10 minutes."

Jumping up from her chair, Zena responds, "You got one and I don't have one. My kids been askin' for 'my daddy.' I say—they say, 'I want my daddy. I want my daddy,' all day."

In the first part of this sequence we again see the complexity in the narrative skills of the two children, especially Zena. We should remember that many people believe that children from disadvantaged backgrounds lack proper language and cognitive skills because of the absence of books or literacy activities in their homes. I have found that for the overwhelming majority of the Head Start children I observed, such a characterization is far off the mark. On the other hand, the children clearly benefit from participating in the Head Start program, which, in terms of quality, provides better care and early education than their parents could afford if they had to rely on private and for-profit child care.

But let's return to the narrative. Zena begins with the use of a rhetorical question ("You know what girl?"), a device she uses throughout her narrative. Rhetorical questions of this type draw attention to statements that follow them and can also serve as topic shifts or extensions. Here, Zena extends the topic of parenting difficulties, noting in effect that her children never seem satisfied. She has given them ice cream and cake, most likely in celebration of a birthday, but they ask for pop (soda), which would add more sugar. Using an internal quote, Zena notes that after her daughter asked for pop "every hour and all day," she told her "No more pop. You're gonna eat ice cream and cake and water—drink water and brush your teeth." After Zena's turn, Debra introduces the idea of a visit to Zena's house, which we see later is a device to set up the discussion of her need for escape from her man "hittin' on" her. Zena immediately responds to Debra's plight but with little comfort, noting that at least Debra has a man, whereas Zena's kids keep asking for their daddy all day long.

While I have no information on domestic abuse in Debra's or Zena's families, several other children in the Head Start center volunteered descriptions of such abuse to the teachers and me over the course of the school year. Although domestic abuse occurs in all social-class groups, what is most important here is how poverty worked against these children's parents' relationships and their family life. Zena's response, that Debra at least has a "man" while her kids constantly ask for their father, is striking. In interviews with Zena's mother I learned that Zena and her younger siblings have been separated from their father for long periods. They stayed in homeless shelters with their mother both before and after this particular role-play episode occurred. Zena's response to Debra clearly shows her understanding of the extent of her mother's (and other single parents') problems in such demanding situations. Facing such challenging family circumstances alone can, at times, be so intolerable that even a mate who is physically abusive might be seen as better than none at all.

The stark difference in the content of these two instances of socio-dramatic play involving upper-middle-class and economically disadvantaged children is readily apparent. For example, the middle-class kids' play addresses the real-life challenges of having to work long hours to run a successful business and the need for charity to help those in need (here, sick kids in the hospital). On the other hand, their role-play also has a number of fantasy elements like rainbow ice cream and ice cream that doesn't melt.

In contrast, Debra and Zena stay very close to the harsh reality of their real lives in the telephone narratives. They talk about and reflect on the difficulties of parenting in poverty. They have no safe parks or reasonably priced grocery stores nearby, and they have to rely on limited and time-consuming public transportation. Most depressing of all is the girls' talk of the absence of their fathers and even domestic abuse. In short, the middle-class children display the joy of fantasy play and optimism about their future as adults, while the Head Start children are much affected by the harsh realities of their family lives and display a sober recognition of what will be challenging futures.

Despite these differences in the quality of life portrayed in the middle-class and economically disadvantaged kids' role-play, their play routines share a number of common features. First, in both examples the kids actively take information from the adult world to create stable and coherent interactive routines in the peer culture. Second, the kids, through their highly sophisticated use of language, embellish the adult models to address both collective and personal concerns in the peer culture. Third, the children's improvised socio-dramatic play contributes to their acquisition of a set of expectations or predispositions through which they confront the circumstances of their daily lives.

While the developing expectations of the upper-middle-class kids are characterized by security and control over their lives, the emerging orientation of the economically disadvantaged kids seems to be one of sober recognition of the difficulty of their circumstances. Yet in both cases, these predispositions are not determined in advance, nor are they simply inculcated by adults. They are, rather, innovative and creative productions in the kids' peer cultures, which, in turn, contribute to the reproduction of the dominant culture with all its strengths and imperfections.

6

"Arriva La Banca"

. . . .

Kids' Secondary Adjustments to Adult Rules

It was a bright, sunny May afternoon in Bologna and I was with three boys who were digging in the outside play area of the preschool. This was my second time doing research at the school. I had spent nine months with the kids and their teachers the year before and now returned for a two-month follow-up. The boys, Alberto, Alessio, and Stefano, were talking about military matters—the navy, warships, and the boss, or *il capo*, on such ships—as they dug holes and buried rocks in the dirt.

I then saw three kids marching around the yard carrying a large red carton. The teachers used the carton to carry play materials to the yard, and I had seen kids play with the carton the previous year. What I didn't know was that the carton was now a forbidden object. I found out later that earlier in the year, before my return, a girl had placed the carton on her head and chased after several other kids. She fell and suffered a minor injury. After this incident, the teachers told the kids that they could no longer play with the carton.

But some of the kids were playing with it today. In fact, they were now marching in my direction and I could begin to make out their chant. It sounded like, *"Arriva la barca! Arriva la barca!"* ("Here comes the boat! Here comes the boat!"). I was not sure about the last word, though; it could have been *"barca"* or *"banca"* (bank). Now the kids were right up close to me. Antonio was leading the way and Luisa and Mario were helping him carry the carton. There was a small blue bucket inside the carton and I could see that it was filled with rocks.

"La barca?" I asked Antonio.

"No, la banca coi soldi!" ("No, the bank with money!"), he said as he cupped his hand in a familiar Italian gesture to stress his correction.

I was fascinated. These kids had created a whole new dimension in banking—a bank that makes house calls! It was a clear advance over the drive-in or walk-up teller.

"Give me some money," I requested.

The kids put the carton down, and Mario took out the bucket with rocks and said, "I'll give him the money."

"How much do you want?" he asked. "There are thousands—."

"Forty thousand," I answered. (This sounds like a lot, but at that time forty thousand lire was only about twenty-five dollars.)

Mario began counting out the rocks, doing exactly as they do in Italian banks by announcing the final sum as he counted out each ten thousand lira note: "Forty thousand, forty thousand, here's forty thousand."

But he gave me only three rocks. "No, no three—thirty thousand. I said forty!"

"Four," Luisa told Mario. "Four!"

Mario then reached in the bucket to get more rocks and counted, "Thirty, forty, here," and handed me three more rocks and then a fourth.

"Sixty now," I said laughing. "Seventy. I said forty!"

"How many?" Mario asked.

Luisa was getting impatient with Mario and seemed to think she could be a better bank teller. "Four, he said four!" she shouted as she tried to take the bucket from Mario.

The three now began to struggle over the bucket, and Antonio scooped the rocks from my hand and dropped them back into the bucket.

"Let's go," he commanded. And the kids marched off again, chanting: "*Arriva la banca! Arriva la banca!*"

I waved and called out, "*Ciao la banca.*"

The kids did not like the adult rule forbidding their playing with the carton, so they played with it anyway, but they created a unique "traveling bank"—an idea taken from the adult world but extended and given new meaning.

The "traveling bank" is an excellent example of children's use of what the sociologist Erving Goffman terms "secondary adjustments." Secondary adjustments involve using legitimate resources in devious ways to get around rules and achieve personal or group needs and wants.

According to Goffman, secondary adjustments are ways in which individuals can stand apart from the self or role that social institutions expect of them. Goffman saw secondary adjustments as forming the under-life of social establishments. Goffman identified a number of types of secondary adjustments in his study of an asylum, a highly restrictive institution. For example, patients had a number of ways of getting around rules about eating food in and taking it from the cafeteria. On days when bananas were served, notes Goffman, "patients would spirit away a cup of milk from the jug meant for those who required milk on their diet, and would cut their bananas up in slices, put on some sugar, and expansively eat a 'proper' dessert." He also observed that on days when portable food, like frankfurters, was served, "some patients would wrap up their food in a paper napkin and then go back for 'seconds,' taking the first serving back to the ward for a night snack."

Goffman believed that his findings in an asylum had implications for understanding the individual's relation to organizations that apply in some ways to all institutions. Furthermore, Goffman argued that individuals have a tendency to both embrace and resist institutional rules and expectations as a way of preserving their personal identity or self.

But is Goffman's work useful for understanding young children's peer cultures? Surely the preschool is not a total institution like an asylum or a prison. Further, preschool children have not fully developed the cognitive skills necessary to define their emerging selves in regard to both their embrace of and resistance to the organizations to which they belong. On the other hand, preschools, like other organizations, have a set of goals, rules, procedures, and expectations for their members. In this sense, Goffman's work has implications for understanding how kids conceptualize and adapt to conventional rules like not bringing personal items from home or moving certain toys from one play area to another (as distinct from moral rules like not hurting another child, lying, stealing, and so on) and procedures of the preschool. And, although preschool children do not have a fully developed sense of self, they do have a clear notion of group identity (kids versus adults). Furthermore, while young children might lack the cognitive skills to infer the implications of both the embrace of and resistance to organizational rules for personal identity, *they do have a clear notion of the importance and restrictiveness of the adult world as compared to children's worlds.* By age two or three, kids have made a distinction between adults ("grown-ups") and kids. In fact, they can, and often do, distinguish between the adult world and their own peer world.

Membership and participation in the adult world are important to kids. Still, children's developing sense of who they are is bolstered by their active resistance to certain adult rules restricting their behavior. In this sense, kids' joint recognition of adult rules, and their common resistance to certain of these rules, can be seen as stable elements of peer culture.

GETTING AROUND ADULT WORLDS:
THE UNDERLIFE OF PRESCHOOLS

Once kids begin to see themselves as part of a group, the mere doing of something forbidden and getting away with it is valued in peer culture. Making faces behind the teacher's back and leaving one's seat or talking during "quiet time" when the teacher leaves the room becomes commonplace over the course of a school term. Even the youngest kids quickly develop an appreciation of these simple secondary adjustments, which can be seen as *exaggerated violation* and *mocking* of school rules. Kids often violate rules solely for the sake of violation— to challenge directly and to mock the authority of the teachers.

Once, in Modena, I observed a group of three- and four-year-olds as they left their main classroom to go to a general meeting room for their optional religion class. The religion teacher began to talk about Jesus and the 12 apostles. Actually she started naming the apostles and had some trouble and asked me for help. As she then proceeded to tell a parable about Jesus and the apostles, one three-year-old girl, Giulia, yawned loudly and said, *"Che annoia!"* ("How boring!"). All the other kids laughed and even the teacher and I smiled at this sophisticated but inappropriate, though honest, commentary. The teacher assured Giulia that the story would get more interesting and continued with the parable.

The mocking of adults or adult control can, at times, be very complex, involving the participation of most, if not all, the kids in the class. In Bologna during the late afternoon the teachers often needed a brief respite before they gave the kids a snack and put things in order for the arrival of parents to pick up their children. To keep the kids occupied during this transition time the teachers often relied on *"disegno libero"* ("free drawing"). The kids sat at tables in groups of five or six and used the magic markers and sheets of paper that the teachers put out. *"Disegno libero"* was a good activity for this time of day. The kids relished the opportunity to have full control over what they drew, and because the activity required minimal supervision, the teachers could relax, talk, and have a coffee.

The volume of noise around the tables was high, but the activity was ordered at first. The children worked on their drawings and the teachers conversed around a table at the far corner of the room near the kitchen. Frequently, however, the loud but consistent hum of activity was disrupted by a dispute. The disputes were usually over the use of particular markers. They often became intense, with children gathering around one table shouting loudly and gesturing. The teachers reluctantly came over and settled things down, pointing out that there were plenty of markers for everyone. But a new dispute soon emerged. A close look at one of these disputes shows that something more complex was going on.

Roberto is looking for a marker that he says, "does not 'write poorly' (*'scrive male'*)." He picks up a red marker, tries it, but tosses it aside, dissatisfied. He then finds another but is again displeased. Now Roberto leaves the table, goes to another, and with the kids at the table not noticing (or pretending not to notice), he takes a red marker. Roberto then returns to his table and begins drawing with the marker. Meanwhile, back at the second table, Antonia rummages through the can of markers and asks, "Where's red?" Maria hands her a red marker, but Antonia waves the offer aside, saying "That one writes poorly." Two other kids now help Antonia and they find several more red markers, but they all "*scrivono male*" ("*write poorly*"). At this point, Antonia slaps her forehead with the palm of her right hand and shouts: "*Ci hanno rubato!*" ("They robbed us!").

This exclamation sets several things in motion simultaneously. Roberto looks up from his work and smiles at the other kids at his table. They all catch his eye and smile back, signaling that they know what is about to happen. Several of the children at the third table look over to Antonia's table and then quickly over to Roberto's. Finally, at

Antonia's table Maria jumps up, points to Roberto and shouts: "*È stato Roberto!*" ("It was Roberto!"). Immediately Antonia, Maria, and several other kids march over to Roberto's table. As they arrive Luisa grabs several markers (including the one Roberto took) and hides them in her lap under the table. Antonia now accuses Roberto of stealing the red marker. He denies it, challenging Antonia and the others to find their precious red marker. As Antonia and Maria begin looking for the marker, Bruna, backed by several other children from the third table, enters the dispute. She claims that Roberto did indeed steal the marker and that Luisa is hiding it. Luisa shouts, "No, it's not true." But Antonia reaches under the table and grabs the markers that Luisa is hiding. Now there is a great deal of shouting, gesturing, pushing, and shoving, and the teachers must again intervene.

I witnessed many recurrences of that event and recorded them in field notes. In fact, this type of dispute occurred, on average, about three times a week in the Bologna preschool, and in all but a few instances it erupted in the afternoon "*disegno libero.*" I concluded that "*Ci hanno rubato*" was really not a dispute over objects, but a mock dispute routine. It was not that there were too few red, green, or whatever color markers that wrote well, but rather that their feigned scarcity enabled the emergence and enactment of the mock dispute. At this time of the day, when the teachers were trying to get the kids to engage in a quiet activity until snack time, the kids would rather argue than draw.

The "*Ci hanno rubato*" routine was a consistent feature of the peer culture and the underlife of the preschool. In the routine, the children challenged adult control (that is, the requirement that they draw to fill time before their snack) and shared a sense of control with each other while they did something they wanted to do (that is, engage in a mock dispute).

In addition to mocking adult authority, preschool children also develop a variety of elaborate strategies to get around rules. Take, for example, rules about the use of objects and space in the preschool. In Berkeley, for example, the children learned early on that certain behaviors could occur in some areas and not in others, and that some play materials were to be used only in the areas in which they were stored and available.

The teachers' concept of space distinguished between inside and outside play. Running, chasing, and shouting were inappropriate behaviors inside the school. This rule was most troublesome for boys, especially the older boys in the afternoon group. Also, the boys in the afternoon group faced an additional rule. Because many of them seldom played indoors during the first month of the school term, the teachers ordered the outside areas "closed" for the first 45 minutes of the afternoon session. The hope was that this rule would prompt the boys to become more involved in indoor play activities.

The rule worked to a certain degree, but it also led to the boys devising a number of ingenious secondary adjustments. One involved several boys' attempts to extend family role-play in the playhouse in interesting directions. For example, the boys proposed that the house was being robbed and took the roles of robbers and police. The police chased the robbers from the house and throughout the inside of the school. When the teachers reminded the boys that there was no running inside, the boys claimed that they needed to run to catch the thieves who robbed the playhouse. Faced with this response, the teacher often compromised and allowed the boys a bit more latitude, but told them to confine the chase to an area near the playhouse. On another occasion, the role-play was extended when a boy suggested that the playhouse was on fire, and, more imaginatively, in a final example, a family was threatened by a wild lion that had escaped from the zoo. In this instance, one boy exuberantly adopted the lion role while another became a lion trainer called in to save the day. In doing so this hero first had to chase the lion all around the school before capturing him to the applause of other kids playing nearby.

These types of secondary adjustments were not confined to boys from the afternoon group. On one occasion, two boys from the morning group, Denny and Martin, and a girl, Leah, were in the upstairs playhouse of the school. They began to cook dinner, but soon became bored with the family play. Denny found a piece of string and, lying flat on the floor, dangled it through the bars (that prevented the children from falling from the upstairs area) and announced: "Hey, I'm fishing. I'll catch us a fish for dinner!" Leah and Martin ran downstairs, got some string, and joined Denny at his fishing hole. Soon other children and the teachers noticed the trio. The teachers were so impressed with this idea that they overlooked the mild violation of proper inside play. In fact, they helped some of the children in the downstairs playhouse to tie toy animals to the dangling fishing lines. Denny, Leah, and Martin ended up catching their limit.

These secondary adjustments are impressive in that they involved the active cooperation of several children, and they can be seen as extensions or elaborations of legitimate behavior for personal ends or goals (for example, the children are allowed to engage in more physical and aggressive play inside the school to catch an escaped lion in pretend play)—what Goffman calls "working the system."

Other types of secondary adjustments involved kids' use of what Goffman has referred to as "make-do's" to get around certain rules. That is, the kids used "available artifacts in a manner and for an end not officially intended." For example, toy weapons were banned from every school in which I observed and there was a general rule against the use of pretend weapons. Yet boys (and a few girls) shot at each other from a distance simply by pointing their fingers and cocking their thumbs. Some kids also converted objects like sticks and broomstick horses into swords or guns or actually constructed weapons with building materials like Legos.

Some examples of "make-do's" were highly ingenious. At the Berkeley preschool, there were several rules regarding the use of play materials. Play with certain toys like blocks, dishes, and toy animals was restricted to the areas where the materials were located. Kids often violated these rules by subterfuge, simply concealing objects on their persons during transport. In one instance, however, a boy, Daniel, took a suitcase from the playhouse, carried it to the block area, and filled it with blocks and toy animals. He then carried the suitcase outside without being noticed, dumped the blocks and animals into the sandpile, and buried them. Shortly thereafter, a teacher noticed the suitcase in the sandpile and told Daniel to return it to the playhouse. He did so without protest, but then quickly returned to the sandpile to play with the blocks and animals. At clean-up time, Daniel abandoned the secretly transported objects and went inside. When a second teacher discovered the objects in the sand during clean-up, she asked two kids in the area how they got outside. They responded with the typical preschool child's answer: "We don't know." In this case this classic excuse was true, but the teacher did not believe them so the innocent kids had to put the toys back in their rightful place.

Kids created many secondary adjustments to get around or delay their required duties at the dreaded "clean-up time." Clean-up time usually occurred at transition points in the preschool day (for example, before snacks or meals, meeting time, nap time, and so on). In all the schools I observed, the general rule was that children stop play when clean-up time is announced verbally or by blinking the lights on and off. The children were then to stop play and help the teachers put the play areas back into order. Many kids questioned the necessity and logic of clean-up time.

At Berkeley one day, clean-up time was announced for the end of the day while I was in the outside sandpile with Peter and Graham, who were filling their dump trucks with sand. Graham tells Peter, "Clean-up time! Ain't that dumb? Clean-up time!"

"Yeah." agrees Peter, "We could just leave our dump trucks here and play with 'em tomorrow."

"Yeah." says Graham as he turns over his truck and shakes out the remaining sand. "Clean-up time is dumb, dumb, dumb!"

Now, a teacher arrives and reminds the boys that they should be cleaning up. They ignore him at first, but after a brief delay they put away their trucks and go inside.

On another occasion a boy, Richard, from the morning group at Berkeley, extended Graham's point about clean-up being dumb by arguing that putting the toys away meant that "we would just have to take 'em out all over again." From the kids' perspective, clean-up is not just work that they don't want to do but also unnecessary. It was dumb work that interfered with fun play.

Given the kids' perception of clean-up, it is not surprising that they devised elaborate strategies to evade it. In the preschools I studied, I discovered a number of categories of clean-up evasion. The first I term the "relocation strategy." When employing this tactic, the kids moved from one area of the school to another immediately upon hearing the clean-up announcement. When asked to clean up in the new area, the kids claimed that they had not been playing there and that they had already cleaned up elsewhere. The teachers soon became wise to this strategy and said everyone had to help clean up wherever they were or whatever they were doing. Although this tactic curtailed the effectiveness of relocation, some children still used it in cases where they had made a big mess of things. They deftly slipped away from the shambles and headed to an area where much less work awaited them. Sometimes this worked and sometimes not, depending on the teachers' awareness of who had been playing where before clean-up was announced.

A second strategy was the "personal problem delay" (claiming you cannot help clean up for one of a number of personal reasons). The kids reported a plethora of problems such as feigned illness or injury ("I got a stomach ache," "I hurt my foot," and so on), pressing business (helping another teacher clean up in another part of the school, needing to go to the bathroom, and so on), or role-play demands (a mother needs to finish feeding her baby, a firefighter has to put out a roaring blaze, and so on).

In one instance in Berkeley, I noticed Brian lying on the ground in the outside yard when clean-up was announced. Shortly thereafter, a teaching assistant, Marie, told Brian to start cleaning up, but he did not respond and continued to lie motionless. Marie then said, "Brian, quit pretending to sleep and start helping us."

Brian still did not move, but Vickie, who was cleaning up, spoke for him. "He can't help. He's dead, killed by poison!" ("Killed by poison!" I love that phrase and will never forget this little drama.)

Marie looked over to me and we both laughed. However, Marie was not about to let Brian escape from work. She knelt down next to him and pretended to pour something into his mouth.

"There," said Marie, "I gave Brian the antidote to the poison. He will now come back to life."

Brian still remained motionless, but I could see a slight smile on his face.

"That antidote didn't work," said Vickie.

"Yes, it did," responded Marie, who began to tickle Brian. Brian giggled and squirmed away from Marie.

"See, he's alive now and is going to help us clean up," said Marie.

Brian jumped to his feet and began cleaning up, not killed by poison after all, but now having less work to do after the long delay.

Sometimes kids try to pull me into their ruses to evade clean-up. Once in Bologna, a girl, Franca, told one of the teachers that she could not clean up because I was in the process of teaching her English. There was some truth in this because children often asked me how to say certain words in English, and Franca and several other children had made such requests earlier in the day. However, we were clearly not involved in an English lesson when clean-up time was announced. Fortunately, I was not brought into this debate because the teacher rejected Franca's excuse out of hand. Nevertheless, during the course of the discussion a good deal of the work of clean-up was performed by other children.

In fact, all of the strategies to avoid clean-up are partially successful for this reason. Because of organizational constraints (that is, teachers' needs to get the children to lunch, to begin meeting time, and so on), any delaying tactic, even a seemingly simple one, is somewhat effective. One of my students, Kathryn Hadley, has volunteered in many preschools and tells the story of a boy, who upon the announcement of clean-up time, went around the school asking teachers and other kids for a "big hug." What a friendly fellow this little "hugger" was at clean-up time. This strategy worked for quite a while before the teachers caught on.

This brings us to a final strategy for evading or delaying the work of clean-up, one so deceptively simple that it took me some time to discern it. It was also highly successful. Let's call it the "pretending not to hear" strategy. When using this strategy, the kids, upon hearing the clean-up time signal, merely continued to play as if the announcement had not been made at all. The teachers then repeated the announcement, usually louder. Still, many children did not respond. The teachers again repeated the command even louder and singled out certain children to begin working. After I first noticed this strategy in Berkeley, I found that it existed in most of the other preschools as well. In Berkeley, I once recorded seven announcements before one group of children responded. All the while, the teachers themselves finished a good bit of the work.

As I mentioned earlier, all of the types of secondary adjustments to evade clean-up can be seen as examples of what Goffman calls "working the system." These secondary adjustments are impressive because, as Goffman notes, "to work a system effectively, one must have an intimate knowledge of it."

But do young children actually share an awareness of these secondary adjustments to the clean-up time rule? In other words, are secondary adjustments really a shared element of peer culture? I believe they are, even though the children seldom discuss secondary adjustments or sit down and plan things out before they act. However, the following two examples (the first from Berkeley and the other from Bologna) support my claim that the kids share an awareness of secondary adjustments.

"Now It's My Turn"

Barbara and Betty are playing in the outside yard near the climbing house. Barbara is swinging on a tire suspended from the roof of the enclosed area of the yard. Betty is standing in front of her, and I'm sitting on the ground nearby. As Barbara swings, Betty bends over and looks down at her and says, "It's clean-up time!" Barbara smiles, ignores Betty, and keeps swinging. Betty now repeats in a louder tone of voice, "It's clean-up time!" Barbara ignores Betty again and keeps smiling and swinging. Betty then repeats "It's clean up time!" seven times. On the seventh repetition Betty raises her voice and draws near Barbara, actually shouting right in her face. Suddenly, Barbara stops swinging, jumps from the tire, and says, "Now it's my turn." "OK," says Betty and she quickly takes Barbara's place on the swing. The routine is repeated with Barbara now shouting, "It's clean-up time."

In this example the kids were actually "playing teacher," with a routine in which the teacher is duped by the kids. In a second example, which I audiotaped in Bologna, a child devises an elaborate scheme for personal reasons. He wants to gain control of a particular object that another child has smuggled into the school (I will talk more about smuggled objects as secondary adjustments later).

"The Scheme"

Felice and Roberto are playing in the outside yard. Felice has a small plastic container that a girl, Angela, has brought to school and given him to play with. It is sort of toy hypodermic syringe (without a needle of course). Before they came outside, the kids have been filling it with water in the bathrooms and squirting one another without the teachers noticing. Once outside, they are soon out of water. Felice shows the container to Roberto, which triggers an idea of bringing water from inside the school to the outside yard. However, the teachers do not allow the obvious transportation of water, for example, in a bucket.

As Felice shows the container to Roberto he says, "Look, I closed it." Roberto then whispers to Felice, "Hey, what if we mix it (the water from the container) with dirt and make a sandcastle? You get water with that. And when you have to go pee-pee you tell—you tell the teacher and she lets you do it. And since you can't always be the one asking, you give it to me and I'll ask. And then 'tum.' I give it to you, and (points to another boy, Armando) he asks, OK?"

Felice, listens to this long plan and merely responds, "Eh?"

"Come on, go and say you have to go pee-pee," Roberto prods impatiently.

"Eh no," says Felice, "not now." And he keeps a close grip on the container.

Roberto, seeing Felice playing with the forbidden object, concocts a highly elaborate scheme to make sandcastles. The idea is not very practical because making enough wet dirt to carry out the plan would require a large amount of water. However, Roberto carefully develops his scheme, whispering and glancing over at the teachers to create a sense of intrigue. Roberto's plan also anticipates the possibility that the teachers might catch on that something is up ("since you can't always be the one asking" to go to the bathroom), and, therefore, includes the devious participation of himself and Armando who was playing nearby.

Finally, the plan involves Felice's giving up the toy container to Roberto, who would then have ample time to inspect and play with it when he goes inside to the bathroom to get water. Because I was right there, I could tell that Felice was tempted by the scheme. However, the requirement to give up the container (which he had patiently coaxed Angela to give him) probably led to his rejection of the plan.

SHARING SUPERMAN: SECONDARY ADJUSTMENTS AND SEEING THE NEED FOR RULES

That example involved a smuggled object. Smuggling is a form of subterfuge common to all the preschools I observed. It was a way to get around the rule, which existed in all the schools, that prohibited (or severely restricted) the bringing of toys or other personal objects from home to school. Children like to bring personal things (especially in their first weeks at the schools) because they provide security in the new environment. Later on, toys and other objects brought from home are valued because they are attractive to other children simply by being different from the everyday play materials in the school.

However, objects brought from home often cause problems for the teachers because the kids fight over them, and a toy might be damaged or broken. Therefore, the rule normally specifies that such objects must not be brought to school and, any that are brought must be stored in one's "cubbie" until the end of the day.

In the American and Italian schools, the children attempted to evade this rule by bringing small objects that they could conceal in their pockets. Particular favorites were small toy animals, little dolls like Polly Pocket," Matchbox racecars, toy soldiers or action figures, and sometimes candy or gum. Kids almost never played with these forbidden objects alone. They immediately sought out a playmate to show their "stashed loot," and tried to share the forbidden objects without catching the attention of the teachers. In Italy, the kids often said to me, "*Guarda, Bill!*" ("Look, Bill") and showed me a toy car or handed me some candy. However, there is something more important than playing with and sharing an object brought from home. The kids feel they are "getting away with something," and in the process breaking down some of the control of the teachers. This shared recognition in the peer culture became as important as having and sharing the forbidden objects.

The teachers were often aware of what was going on, but simply ignored minor transgressions, overlooking these violations because the nature of the secondary adjustment often eliminated the organizational need to enforce the rule. Kids shared and played with smuggled personal objects surreptitiously, to avoid detection by the teachers. If the kids always played with personal objects in this way, there was no conflict and hence no need for the rule. Thus, in an indirect way the secondary adjustment endorsed the organizational need for the rule.

These examples now bring us to a recognition of how secondary adjustments can actually help kids understand the need for certain conventional rules, and how, in addition, they influence teachers to modify their definition and enforcement of rules.

Let's begin with the former; an example from Bologna is helpful. One day a girl, Luisa, brought a small plastic replica of Superman in her pocket to the preschool. At one point, she took out the toy to show it to a boy, Franco, with whom she was trying to cultivate a special relationship.

As Franco ran by with some other boys, Luisa held up the toy and said, "Look, Franco it's Superman."

"Hey, beautiful," said Franco as he took the Superman and began flying him around.

The two played nicely with Superman, passing him back and forth without incident for more than half and hour. However, at one point, Luisa complained that Franco was getting Superman dirty, keeping the toy in his possession too long, and not sharing properly. Franco dismissed these complaints and continued playing with the toy.

Luisa said, "Give it back or I'll tell the teachers."

Franco ignored this threat and Luisa began walking over toward where two teachers were sitting. I could see that they had not noticed Luisa and Franco playing with Superman or were unconcerned because there was no problem. After Luisa got about halfway to the teachers she stopped, waited for a few seconds, and then walked back toward Franco.

Luisa now found herself in a quandary. She realized that if she complained to the teachers about Franco's refusal to share the toy, she herself would be reprimanded for bringing Superman to school in the first place. In fact, Superman would probably end up in one of the teachers' pocket until the end of the day. Therefore, Luisa waited patiently and when she saw her opportunity she grabbed the Superman from Franco, saying, "Basta!" ("Enough") and put the toy back in her pocket. Franco protested and even threatened to go to the teachers. He ran in that direction, but soon veered off to join the boys he had played with earlier. He also knew that the teachers would not get Superman back for him. So he was off, leaving Luisa alone with her Superman.

SHARING DAYS: SECONDARY ADJUSTMENTS AND REPRODUCTION AND CHANGE IN ADULT CULTURE

Clearly, in trying to get around the rules the children are beginning to understand why the rules exist. But how do secondary adjustments affect teachers' reevaluation and even changing of their own rules? To better understand this, let's return to the Bolognese kids' "traveling bank" that I discussed in the opening of this chapter.

As we know, the carton the kids are using is a forbidden object, but the teachers allow the activity of the "traveling bank" to continue because of its creativity. However, after the kids take care of my banking needs and continue to march around chanting "Arriva la banca," Antonio and Mario begin to struggle over possession of the carton. At this point a teacher intervenes.

As she approaches the kids, the teacher asks, "One little girl already ended up in the hospital because of the carton, do you remember?"

Luisa shakes her head yes, but the boys do not respond and Mario has started to cry.

"Right, Carla," says the teacher referring to the girl who was hurt playing with the carton. "What happened? You were crying, what happened?"

"Well, I was the one giving the carton," says Antonio.

"You were crying because he didn't give you the carton or because he hurt you?" the teacher asks Mario.

"Because—," starts Mario.

"Why didn't he give you the carton?" the teacher interrupts.

"Hey, why do you need to cry?" she asks Mario. Then waiting a second she says to Antonio, "And you, why are you being a bully?"

Mario has stopped crying now, and he and Antonio stand with their heads down, having been chastised.

The teacher now returns to her talk about what happened to Carla. "This carton here," she says, pointing to the edge of the carton, "do you remember when Carlina—the mark that she has here?" The teacher now points to a place on her forehead where Carla sustained her injury. "Carla, Carla come here," the teacher calls out for the child, who is playing in another part of the outside yard. "Cause Bill was not here and he does not know," says the teacher.

Carla now arrives, and then the teacher turns to me, and says, "Carla because of this," she taps the carton, "to the hospital."

"When," I ask.

"Eh—in—Septem—October, when you weren't here," the teacher replies.

"Yes," I say.

"Because they had pushed like this," the teacher points out and displays what happened by moving the carton to Carla's forehead.

In this sequence the teacher is encouraging the children to link a past event (Carla's injury while playing with the carton) to their present behavior (fighting over the carton). This is the first phase of a complex intervention strategy in which the teacher indirectly evokes the rule that forbids playing with the carton. The teacher summarizes the past event, stating the outcome (a little girl ended up in the hospital) and the cause (playing with the carton), and ends her turn with the tag, "Do you remember?" The children are now primed for a series of more direct questions about the event and its relevance to the rule. The teacher begins with the identification of the injured girl and then asks what happened to her. Luisa signals that she remembers, but the boys are still embroiled in their dispute about control of the carton and Mario starts to cry. The teacher tries to get to the bottom of why Mario is crying, but decides it is not a serious matter. She playfully chides Mario for his oversensitivity and warns Antonio about being a bully.

Having now gained the boys' attention, the teacher returns to the rule about the carton. She picks up the carton and asks if the children remember the scar that Carla acquired because of her accident. Before the children can answer, the teacher calls for Carla to join the group. While we wait for Carla, the teacher reminds the children that I was not present when the accident occurred and she fills me in on what happened. Although we cannot be sure if the teacher's elaborate recreation of the event is primarily for me or the children, it is clear from what happens next that the children are collectively reliving the experience of Carla's injury.

The teacher now continues to talk about what happened, "I knew—."

"Who did it?" interrupts Antonio (meaning who was responsible for Carla's injury).

"I know who she was," the teacher says smiling. "We don't say who did it."

The teacher now taps Mario, Antonio, and Luisa on the head, and says, "So, we don't play with the carton, it's an awful toy."

"Here," says Mario and puts some rocks in the bucket.

"Play with it, using it like this," says the teacher as she places the carton on the ground indicating that they can play with it only in this manner.

The kids leave the carton on the ground and place the bucket in it. They continue to play as if they are bankers for a while longer without moving the carton. Before long it is time to go inside for a snack.

In the last part of the teacher's discussion with the kids, Antonio's interruption to ask "Who did it?" displays a curiosity typical in kids' peer culture: the desire to have information about the mischief of a playmate. The teacher's response is composed of two contrasting declarative sentences. In the first she states emphatically that she knows who it was. In the second sentence she contrasts this emphasis with the general rule of "not telling on others" or "dwelling on the past misbehavior of others" ("We don't say who did it").

The teacher goes on to provide the moral of the re-creation of the past event: "So, we don't play with the carton, it's an awful toy." Nevertheless, she tells the kids they can play with the carton, but only if they keep it on the ground and use it as a container.

Underlying the teacher's actions is a philosophy that stresses the rights and welfare of the group. This philosophy is manifest in several ways. First, when the kids began playing with the carton, the teacher decided to relax the normal restriction of play with a potentially dangerous object and to monitor the activity from a distance. Thus, the creative use of the carton temporarily suspended its threat to the general welfare of the children. However, once a struggle over the carton ensued, its potential danger reappeared and the teacher intervened in the play. The teacher's hesitancy in enforcing the rule displays both awareness and appreciation of the creativity and autonomy of communal aspects of peer culture.

Second, when the teacher intervened, she did not immediately enforce the rule, but rather subtly drew the kids' attention to the reason for its existence. She did this by encouraging the communal re-creation of the event that brought about the establishment of the rule. This re-creation involved questioning the children, providing me with information, and even examining the scar on Carla's forehead.

Third, the teacher resisted Antonio's request that she identify the child responsible for the earlier accident. This resistance displays her emphasis on the importance of the rule for the general welfare of the group, as opposed to shaming a particular child for misbehavior. In short, her message is not that "so-and-so's behavior was bad" and that "you (as an individual child) should not repeat it," but rather that "what happened to Carla could happen to any member of the group, so we should all be careful when playing with the carton."

This incident of the Bologna teacher relaxing rules in response to creative reproductions was something I also observed in other schools. In some American preschools, the teachers' response to kids' smuggled objects was to proclaim a "sharing day." On that day, children were encouraged to bring personal objects from home. The kids then displayed and described their toys and other personal possessions in a "show and tell" routine at circle time. Later, during free play, the children shared their possessions with friends. Thus, the teachers built on the kids' secondary adjustments (desire to bring personal objects) in order to create an activity that stimulates the children's development of communicative skills and social knowledge. Simultaneously, the teachers controlled the disruptive potential of the personal objects through the idea of a formal sharing period in free play.

This change in the school curriculum did, however, have some negative effects. Over time the kids began to evaluate both the objects and their peers' presentations during "show and tell." First, they became aware of the range of their peers' possessions; an increasing consumerism developed in the peer culture, which was passed on quickly to parents who often complied with their children's requests to buy a new toy for sharing day. Second, some children's possessions and performances were devalued ("Oh, no, Jenny brought that old bear again!"). This situation contributed to a status hierarchy that was related, at least in some way, to the parents' economic resources.

Overall, these examples and discussion display the complexity of cultural contact between the adults' and kids' worlds. These contacts can be cooperative, harmonious, constructive, and negative. Regardless, kids' secondary adjustments and adults' reactions to them bring about social change in the kids' and adults' cultures.

7

"You Can't Come to My Birthday Party"

· ·

Conflict in Kids' Culture

During clean-up time in one of the Bloomington preschools, I watch three boys (Martin, Jason, and Bill) chide Mickey for not helping. Mickey gets upset and says he will not be their friend anymore. The others ignore this threat and continue cleaning up. Mickey stomps off and sits at a nearby table. He seems to realize his error and starts to cry, saying "I will not have any friends anymore." Bill and Jason try to comfort him and say they still like him, but are just mad at him for not helping. Mickey is still crying some, but is cheered up and when I ask if he is OK, he says "yes." With clean-up finished, the boys go to get books to read before going outside.

Here we see Mickey's attempt to use the "denial of friendship" strategy to get the upper hand in a dispute. When ignored, he realizes that his threat, if carried out, could leave him without friends. Mickey is now more upset than he was in the original conflict. However, he is comforted by his friends who say they still like him, but were mad at him for not helping.

Although conflict among kids is often troubling and annoying to adults, it is a natural element of children's culture and peer relations.

Comparative research of kids' cultures shows that conflict contributes to the social organization of peer groups, the development and strengthening of friendship bonds, the reaffirmation of cultural values, and the individual development and display of self. Many of these, more positive, elements of conflict can be seen in the example given.

Although children's interactions are generally harmonious, conflicts and disputes are far from rare. Kids get into disputes over possession or control of play materials, the general nature of play, access or entry into play (including friendship disputes), verbal claims, accidental injury, and in response to unprovoked aggression. Most disputes are short lived because adult caretakers usually intervene quickly. In fact, the sociologist M.P. Baumgartner has argued that children seldom negotiate compromises in their conflicts because, like those in all subordinate populations, they quickly cede control of their disputes to those with greater power and authority. However, children's conflicts and disputes across cultural and subcultural groups vary considerably and, in groups where kids are given more opportunity to settle their own conflicts, highly complex negotiated settlements occur. Let's take a comparative look at some of my research on conflict and disputes among white middle-class American, African-American, and Italian children.

"I HAD IT FIRST": CONFLICT AND DISPUTES AMONG WHITE MIDDLE-CLASS AMERICAN CHILDREN

Among the white middle-class American kids I studied, the most frequent types of conflict related to the nature of play and disputes over objects. These disputes were usually short and confined to two or three children. More elaborate, serious, and emotional disputes usually centered around friendship. Disputes over the nature of play were frequent and varied. They were often quite simple in structure.

Three girls (Alice, Beth, and Vickie) are playing a card game in the Berkeley preschool. Alice and Beth try to take a turn at the same time.

"No–o-o-o!" says Beth.

"It's my turn," counters Alice.

"It's her turn," confirms Vickie.

"My turn. It's my turn!" insists Alice.

"Beth, it's her turn," Vickie repeats.

Beth, however, simply ignores the other girls and takes a turn anyway. Then Alice takes a turn and the play continues without further conflict.

This example is fairly simple in structure in that it does not really get beyond claim and counterclaim. Some disputes over the nature of play can, however, become quite complex as the kids come up with some innovative reasons for their positions in a dispute.

In the Berkeley preschool, Rita, Denny, and Martin have created a role-play scenario in which Rita is Denny's mother and Martin seems to be Denny's friend. Rita pretends to make a pair of pants for Denny while he plays on the climbing bars with Martin. All three are on the bars, but Denny now climbs higher and leans out, holding on with one hand.

"Get off the porch! It's dangerous," warns Rita.

Denny keeps climbing and ignores Rita.

"Get off that porch. Get off it. It's dangerous. You'll fall off it," Rita scolds Denny.

"OK. Then make him get off the porch too," says Denny pointing to Martin.

"Cause my brother did one day," says Rita seemingly referring to an accident her real brother had.

"Well, you have to share," says Denny.

"I am sharing," counters Rita. "But you're like a monkey just hanging on the porch and you're gonna fall off."

This dispute begins with Rita, in line with her role as mother, objecting to Denny's dangerous play. Denny protests, arguing that Martin must also obey. Rita ignores this ploy to include Martin and offers an additional and more specific reason for why Denny should obey: Her brother once fell off the porch. Whether this mishap actually occurred is impossible to know, but Rita is clearly going outside the present role-play to provide support for her position. Denny then makes a reference to the general rule of sharing, which is somewhat strange given that they are already sharing and playing together. In effect, Rita makes this same point when she argues that she is sharing, and then she goes on to repeat her reason for opposing Denny's behavior (that is, it is dangerous). This example nicely demonstrates the functions that disputes can perform in organizing and embellishing role-play events.

Object disputes normally had a simple structure of opposition-reaction that could be repeated over and over again without elaboration.

Barbara and Richard are playing in the block area of the Berkeley preschool. Richard picks up a block near Barbara and Barbara tries to grab it from him.

"No. No!" says Barbara, struggling with Richard for the block.

"No," says Richard.

"I had it first!" counters Barbara.

"I want one."

"But I had it first!"

"I want one, Barbara."

"I had that first."

"I want it."

"I had it first!" Barbara shouts.

At this point a teacher intervenes and suggests that Barbara go to the shelf and get another block. As Barbara does so, Richard takes the disputed block. Barbara returns with her block, but does not seem happy with the outcome.

Sometimes possession disputes and disputes over access to play areas can become serious, so that teachers are quick to intervene, but in some instances children solve things on their own, using humor to relieve the escalating tension.

Richard and Denny are playing with a slinky on the stairway leading to the upstairs playhouse of the Berkeley preschool. Joseph and Martin enter and stand near the bottom of the stairs.

"Go," yells Denny.

Martin runs off, but Joseph moves halfway up the stairs and, pointing to his shoes, says, "These are big shoes."

"I'll punch him right in the eye," Richard says to Denny.

"I'll punch you right in the nose," Joseph responds.

"I'll punch him with my big fist," Denny says to Richard.

Joseph tries to respond, "I'll—I—I—," but he is interrupted.

"And he'll be bumpety, bumpety and punched out all the way down the stairs," says Richard.

"I—I—I'll—I could poke your eyes out with my gun. I have a gun," says Joseph.

Denny replies, "A gun! I'll—I—I—even if—."

"I have a gun too!" Richard declares.

"And I have guns too," says Denny, "and it's bigger than yours and it poo-poo down. That's poo-poo."

All three boys laugh at Denny's reference to poo-poo.

"Now leave," says Richard.

"Un-huh," says Joseph. "I'm gonna tell you to put on—on the gun on your hair and the poop will come right out on his face."

"Well," says Denny.

"Slinky will snap right on your face too!" challenges Richard.

Denny adds, "And my gun will snap right—."

At this point Debbie enters and says she is Batgirl. She asks if the boys have seen Robin. Joseph says he is Robin, but Debbie says she is looking for a different Robin and runs off. Then Denny and Richard move from the stairs up into the playhouse and Joseph follows them. From this point until clean-up time the three boys play together.

In this example, Joseph's standing up to Richard and Denny's threats is eventually successful in his gaining access to the group. In fact, his refusal to back down leads to a series of escalating threats that would bother most adults. In the sequence, Richard and Denny work together, supporting one another indirectly by *telling each other* what they will do to Joseph. This team effort leads to Richard's creative simile in which he states that Joseph will go "bumpety, bumpety and punched out all the way down the stairs" like the slinky they were playing with earlier. Joseph is not intimated by this threat and escalates the dispute, saying that he could poke Richard's eyes out with his gun.

In the midst of these threats, Denny introduces a humorous tone with the mention of poo-poo. Joseph picks up on this, noting that his gun will put poop right on Richard's face. The boys then return to physical threats, but Debbie's entry quiets the threats. She says she is looking for Robin and does not disagree with Joseph when he says he is Robin. Instead, she says she is looking for another Robin and leaves the scene as quickly as she entered. With Debbie's departure, Joseph follows Richard and Denny up the stairs and is accepted into their play.

While Debbie avoids direct conflict, Joseph, by contrast, seems to initiate a fight when the other two boys attempt to exclude him. In the end, the altercation does not prevent the boys from playing together, but in fact facilitates it. Although it seems surprising that the boys play together after such angry threats, Joseph's persistence and the use of humor to relieve the tension are important elements in the resolution of the dispute. As we discussed earlier, "showing you can play" is essential for gaining entry. For boys' play, which is often rough, being tough and standing one's ground in threats and insults can lead to affiliation.

In the examples we looked at so far, the white middle-class American kids play with language, develop some logic skills in making and defending their positions, and manage to keep disputes from becoming too emotional. Thus, we see that mild conflict and disputes can play positive roles in peer culture. As we saw in Chapter 3 and in the example at the start of this chapter, kids often have disputes about, or tie disputes to, friendship. Often, these disputes can become emotional, especially when children who consider themselves best friends are involved. In other instances, the conflicts are less intense because the children try to manipulate or control the behavior of their friends rather than directly challenging friendships.

Three five-year-old girls, Ruth, Shirley, and Vickie, are sitting at a table in one of the private preschools I studied in Bloomington. They are looking through department-store catalogues and selecting items to cut out and paste on paper to make a collage. The girls have decided to concentrate on items they refer to as "girls' stuff," referring to some other items as "yucky boys' stuff." Shortly after the activity is under way, another girl, Peggy, comes over to the table and stands near Shirley.

"We don't want that couch. That's dumb," says Shirley, referring to a picture of a couch in one of the catalogues.

"All we want is the pretty stuff," says Ruth.

Peggy now speaks for the first time: "If you are going to come to my birthday, you have to obey my orders."

"Oh," replies Ruth. "We don't care."

Ignoring Ruth's rebuff, Peggy continues: "And every girl in the *whole* school is invited. Shirley, *every* girl. I'm goin' to put a sign that says: 'No boys allowed!'"

"Oh good, good, good! I hate boys," Vickie responds.

"And the girls can't do whatever they do—they gotta obey my rules," declares Peggy.

"Then we're not coming," snaps Shirley.

"Yeah, but the point is," says Ruth. "Wer'e cutting out all these things for, for presents for your birthday, but we'll forget about it. We're not coming!"

Peggy does not respond to this last retort and there is a short silence and then Ruth says, "I got a Christmas wish book, and I cut out a whole bunch of stuff for my birthday."

"Me too," says Vickie.

Shirley is flipping through pages of one of the catalogues and notices items for males. "Oh gross! That's for boys. Uhh, boys' stuff. Yuck!"

"Oh, well. Don't look at the boys' stuff," says Vickie.

"Oh. Look at the cute little bunny rabbit," says Shirley in an endearing voice.

Now Peggy walks away from the table without saying anything. The other girls do not seem to notice and continue to talk and cut out items from the catalogue for about 10 more minutes until "clean-up time" is announced.

In this complex episode, the four girls take the opportunity of looking through the catalogues and cutting out pictures to reaffirm gender role stereotypes. In fact, the girls build a good deal of solidarity in their praise of girls' stuff and their condemnation of "yucky" boys' stuff. However, what Peggy is up to is not so straightforward. She enters the conversation by noting that if the other girls are going to come to her birthday party, they have to obey her orders. It is unlikely that the reference to her birthday and obeying her orders is an attempt to join the activity, because there is an empty chair at the table. Also the other girls make no attempt to resist her presence in the area. Finally, it would be an odd strategy (especially for a girl) to gain access by picking a fight.

So what's up with Peggy? Peggy's reference to her birthday party is most likely an attempt to gain the attention of Vickie and Shirley, with whom she often plays and claims best-friend status. This interpretation is supported by the fact that Ruth (a nonmember of the friendship group) first rejects Peggy's attempt at social control by noting "We, don't care" about the birthday party.

Peggy ignores Ruth and expands her discussion of her birthday party to an exclusion of boys—only girls will be invited and she will even put up a sign that says "No boys allowed!" This strategy ties in nicely with the other girls' disparaging of boys' items in the catalogue. Vickie enthusiastically supports the exclusion of boys. Now with some support, Peggy sets another edict for the planned party: "And the girls can't do whatever they do—they gotta obey my rules." Peggy is clearly pushing things to the limit here and, friend or no friend, Shirley is having none of it. She dismisses the whole business with a simple "Then, we're not coming." Ruth quickly jumps at her chance to build good relations with Shirley and Vickie noting, "Yeah, but the point is, we're cutting out all these things for your birthday, but we'll forget about it. We're not coming!"

Of course, the girls had not discussed cutting out things for Peggy's party. But Ruth's claim fits the ongoing discourse (after all they could have been) and it totally defeats Peggy's attempt to insert herself into the play. Sensing her defeat, Peggy soon withdraws from the interaction.

Although the girls seem to take Peggy's proposal about her birthday party seriously, it is doubtful that they really believe that a birthday party for only girls in the school will ever take place. It was common practice in the school that all children were invited to birthday parties. In fact, Peggy had her birthday party, to which all the kids were invited, two months earlier. These facts suggest that the conversation is primarily about the nature of friendship relations among these particular girls. Peggy sees her friendship with Shirley and Vickie threatened by Ruth. So she tries to create control—nicely building on the "We don't like boys" theme, but going too far. In the end, instead of increasing her solidarity in the group, she creates a rift in which she is rejected (at least for the moment), and Ruth becomes more actively involved. Much like the case with Mickey that we saw in the start of this chapter, attempting to control your friends can lead to conflicts that threaten enduring affiliation.

"JESUS IS BIGGER THAN EVERYBODY": CONFLICT AND DISPUTES AMONG AFRICAN-AMERICAN CHILDREN

From the perspective of white middle-class adults, the African-American kids I observed in the Indianapolis Head Start center would probably be somewhat rough and threatening. The kids were more likely than the white middle-class American children I studied to resort to pushing and shoving in disputes over objects or, more frequently, to get a place near the front of the line going to the bathrooms or outside to play. The teachers tolerated rough behavior to some degree, saying it did not really matter who was first in line once we got to the bathrooms or outside. However, when the kids were too rough, the teachers intervened and scolded them and threatened to tell their parents about their misbehavior. Such threats were taken seriously by the children, who showed a great deal of respect for the teachers' authority.

Also as we discussed in Chapter 1, the African-American kids frequently engaged in "oppositional talk." This type of competitive talk and joking was sometimes produced in sessions of ritualistic teasing and insults, similar in structure to those produced by African-American preadolescents and adolescents. In other instances, teasing and oppositional talk were embedded in disputes over the nature of play or possession of play materials. Here's an example regarding the nature of play.

Pam and Brenda are playing in the sandbox making sand pies and cakes at the Indianapolis Head Start center.

"Hey girl," Pam tells Brenda. "Don't use that little ol' thing [small scoop]. Use this big one!"

"OK," says Brenda, taking the bigger scoop.

The girls place sand in various pots and pans for a while and then Brenda says: "What's a matter with you girl? That's too much sugar in that cake!"

"No, it ain't." says Pam.

"I said it is, girl," Brenda replies.

The play continues with this sort of back-and-forth evaluating and teasing for about 20 minutes. Neither girl seems to dominate the other, and neither gets offended or upset. In fact, the conflict and teasing spices up the play and makes it more enjoyable.

In a second example, involving the only two African-American boys at the Berkeley preschool, a dispute is produced in this same oppositional style. The dispute is more serious and involves possession and use of play materials.

Daniel and Tommy are hammering nails into boards. They are standing on chairs and working with the hammers and boards on top of a shelf. They are preparing for the "Almost Puppet Show" we discussed in Chapter 4. I'm sitting with several other children in front of the shelf, waiting for the puppet show to start. Daniel leaves briefly. When he returns, he sees one board on the floor and Tommy still hammering the other board on the shelf.

"Why did you have to—," says Daniel, then he stops and looks down at the board on the floor. "Hey, where's my board? Tommy, this is my board," says Daniel as he grabs the board Tommy is working with. "Go get your own."

Tommy looks down and sees the other board on the floor. "My board's right down on the floor?" asks Tommy. "That's your board!"

Now both boys get down and inspect the board on the floor.

Tommy picks up the board from the floor and says, "I wasn't working this."

"You weren't? Then who was?" asks Daniel.

"I was working yours?" asks Tommy. "Not then. I was working—."

"That one's mine!" insists Daniel. "I was working on there and that was mine, huh?"

Now the boys get back up on the chairs and Daniel pulls the board on the shelf in front of him. "That was mine. Here. You go get your own. That was mine, so get your own."

Tommy gets back down and inspects the board on the floor. He shakes his head, but decides to work with it.

"Good," says Daniel.

Daniel negatively reacts to Tommy using the remaining board on the shelf by claiming that the board is his. He tells Tommy to get his own board, which is on the floor. Tommy denies that it is his board, but not in a simple way (for example, "It is not!"). Instead, he uses a rhetorical question: "My board's right down on the floor?" He then follows with a denial: "That's your board!" Tommy's turn at talk here is highly stylized and the first part (the rhetorical question) is what the anthropologist, Marjorie Goodwin, calls a "predisagreement." In other words it is an element of speech that precedes and signals the coming disagreement, which in this case is: "That's your board." This stylized way of disagreeing is not only complex, it spices up disputes and conflicts—in a way extending, or what Goodwin calls "aggravating," them. Goodwin found that aggravated disputes were common among black preadolescents.

Daniel picks up on this stylized talk with a predisagreement of his own—"You weren't?" (another rhetorical question)—followed by a challenge, "Then who was?" This challenge is a counter to Tommy's denial because, if Tommy was not working with the board on the floor, who else could be except Daniel, who already denied that the board on the floor was his? Tommy responds with another predisagreement ("I was working yours?"), followed immediately by a denial and what seems to be a reason. However, Daniel interrupts Tommy before the reason can be stated and first repeats his claim to the board on the shelf and then supports it with a reason produced in the form of a tag question ("I was working on there and that was mine, huh?"). Tag questions contrast with predisagreements, which are forward-looking and send the message "I am going to disagree with you," in that the tag ("Huh?" "OK?" or the Italian *Capito?* ("Understood?"; see later discussion) instructs the listener to reflect back on what has just been said because it is obviously correct. At this point, Tommy seems to tire of the debate and agrees to use the board on the floor.

In addition to producing stylized oppositional talk in brief exchanges and disputes, the African-American kids also engaged in extended group debates. These debates often grew out of conflict resulting from one or more kids opposing the stated beliefs or opinions of another kid.

Although the source of these group debates was often related to competitive relations among the African-American kids, the debates themselves revealed much about the children's knowledge of the world and served as arenas for displaying self and building group solidarity.

In the Indianapolis Head Start center, several kids (Roger, Jerome, Darren, Andre, Ryan, Alysha, and Zena) are at the same table eating lunch. I am sitting at the table having lunch with them and the teacher is sitting nearby at the serving table. The rest of the class is having lunch at two nearby tables. Roger and Jerome are good friends and value frequent competitive talk about their knowledge, skills, and possessions. It is such talk that sets off the following group discussion.

"I saw somebody on *Hard Copy* [a television show] who had a bullet through the back of his head," says Roger.

"I'm getting—I'm getting hard copy in the back of my head," Jerome replies.

"You can't get that word in the back of your head," counters Roger.

"OK [inaudible] in the back of my head," says Jerome.

"Can't get that word either."

"Yes, I can."

"Un-uh."

During these exchanges there is quite a bit of other talk going on at the table among the other children. Thus, it is difficult to transcribe the first part of Roger and Jerome's discussion, which I videotaped. Roger begins by talking about a show that he watches, *Hard Copy*, a type of TV tabloid show, which was popular when this conversation occurred. Jerome's reply that "I'm getting hard copy on the back of head" is clearly a way to start a dispute with Roger. It might seem far-fetched, but at that time it was the style in the African-American community in which these children lived for young males to have certain words (usually their names or nicknames) carved in the back of their heads when they got haircuts. In fact, Jerome had his nickname carved in the back of his head. Roger, however, denies Jerome's assertion on the grounds that it is too long a phrase to be carved into a haircut. Jerome then proposes another, inaudible, word to be carved in, but Roger rejects that word as well. At this point, the boys start talking about other programs they watch, and Jerome names a show that was again inaudible because of noise at the table. Things then quieted down a bit and Jerome and Roger were the only ones talking. They continued their dispute, but soon after, other children joined in.

"It [referring to a television show] comes on every night," says Jerome.

"We watch that channel, and it don't come on our TV," says Roger. "We got 80 channels. And we got that channel, but when we watch that channel, that don't even come on. What channel it come on?"

"HBO," answers Jerome.

"We watch HBO," says Roger.

"It comes on cable," says Jerome.

"We got cable," declares Roger.

"We got cable too. For real," Zena says.

"We do too," says Ryan.

"We do too," adds Darren.

In this part of the discussion, Jerome and Roger turn to a debate about what programs they receive on their televisions. This shift in topic moves the discourse from general competitive talk to a debate about a specific issue that can be factually settled. Roger moves in this direction when he argues that they receive 80 channels (an obvious exaggeration at the time this debate occurred) on their television; therefore, it does not seem likely that the program Jerome is talking about is really on television. Roger ends his turn with a challenge to Jerome to name the channel. Jerome says it comes on HBO. Roger counters by saying that they watch HBO. His point is that the program is not on that channel because if it were, they would have seen it. Jerome now says that the program comes on cable, but because HBO is only on cable, either this claim is redundant or Jerome is confused. Roger again counters with a simple declarative that they have cable. In this exchange, Roger continually opposes Jerome's claims with appeals to fact and logic, displaying the cognitive complexity that such competitive talk can generate.

Jerome can escape Roger's interrogation because the mention of cable moves the competitive talk to a group discussion. Several children (Zena, Ryan, and Darren) now enter the talk and all note that they have cable. Zena's entry into the discussion is especially interesting because she says, "We got cable. For real." Zena is from a very poor family, so poor in fact that she, her mother, and sisters were living in a homeless shelter. However, the shelter had a hookup to cable television, which is probably why Zena thought it was necessary to add "for real" to her claim. Here we see how the children's personal histories and experiences become part of group debates. The discussion now continues with a debate about who has the biggest cable.

"I got the biggest cable. I got the biggest cable," challenges Roger.

"I thought all cable was the same," says the teacher.

"So did I," I say, laughing.

"They ain't either," says Jerome. He then puts one of his hands slightly under the table and the other up above his head and says, "My cable's this big!"

"Un-uh," denies Zena.

"My cable's 'bout this big," says Roger, as he holds his hands about two feet apart.

"Jesus is bigger than everybody," Alysha says, very softly.

"My—my cable's like this big," says Darren, holding his hand about two feet above the table.

"Marvin's head is bigger than anybody's," says Zena, teasing a boy at another table.

"I'm bigger than Jesus," says Jerome, responding to Alysha.

"Nah-uh," says Alysha. "Jesus is bigger than everybody!"

"My cousin's bigger than Jesus. My cousin is that big," says Jerome, as he holds his hands far apart.

"But he don't do—this," says Alysha, as she reaches her hand up as far as she can from the table. "He's [Jesus] this big."

"My cousin's this big," says Jerome, as he raises his hand higher than Alysha's.

"Alysha," says the teacher, "get through so you can drink your milk today."

"He's this big." says Andre, speaking for the first time. He holds his hand up higher than Jerome did.

"Who? Who?" asks Jerome.

"Jesus," Andre answers.

The talk about cable television leads Roger to claim that he has the "biggest cable," prompting the teacher and me to remark that cables are all the same size. However, Jerome denies the adults' contention, and the talk about the size of cable continues, with the kids in firm control of their dispute. After Jerome and Roger say and demonstrate with their hands how big their cable is, Alysha speaks for the first time.

She builds on the competitive talk about cable size to argue that "Jesus is bigger than everybody."

Alysha speaks very softly and it is not clear that others heard her until Jerome claims, after several other kids have spoken, that he is bigger than Jesus. Alysha comes from a large family and from interviews with her mother, I know that they do not have cable television in their home. In fact, Alysha's family was one of the most economically disadvantaged in the Head Start program. Both her parents were in the home and her father worked. However, his income was meager and there were six children in the family, the oldest only seven years of age.

Not having cable, Alysha remained quiet during the debate about watching television shows on cable. However, when the discussion turned to size and who had the biggest cable, she saw her chance to participate and seized it. Alysha's family is very religious. Her father worked at a religious radio station and the family attended religious services several times a week. Alysha's mother and father were also very active in the church, holding a number of important and time-consuming positions. So, when the discussion about what is the biggest came up, Alysha, relying on her religious training, said softly but firmly, "Jesus is bigger than everybody."

Several other children were talking at the same time Alysha made her claim. Darren, responding to Jerome and Roger, moved his hand above the table and said his cable "is like this big." Zena tried to use the talk about things being big to tease Marvin, who was eating at the next table, saying that his "head is bigger than anybody's." This type of banter across tables was frequent during lunch, but Marvin ignored Zena and the playful teasing stopped there. At this point, things quieted down and Jerome challenged Alysha, saying "I'm bigger than Jesus." This challenge confirmed Alysha's entry into the debate, and Alysha immediately repeated her earlier assertion that Jesus is bigger than everybody.

Alysha's claim was related to her religious beliefs that Jesus is all-knowing and all-powerful. At that point, the mainly jocular discussion of the size of cable became more serious. However, Jerome's assertion that his cousin is bigger than Jesus was clearly presented in a non-serious way. Alysha stayed in the more serious vein and was wisely doubtful about Jerome's cousin, arguing that the cousin could not reach up as high as Jesus is tall.

The teacher then demanded that Alysha make progress on her lunch and, thus, drew her away from the discussion. However, Andre, speaking for the first time, took up Alysha's position, asserting that Jesus is indeed the biggest. The discussion ended soon thereafter, when the teacher told the children to begin to clean up their places and get ready to brush their teeth.

This example is representative of the types of competitive peer talk that occurred routinely at the Head Start center. Participation in such competitive talk builds a general peer-group identity and, at the same time, provides the children with opportunities to display their personal knowledge and interests. Overall, oppositional talk, which many white middle-class adults might see as negative or hurtful, has many positive features in the Head Start peer culture. Oppositional talk (in short, dyadic exchanges and extended group debates) drama-tizes everyday interaction and provides the kids with an arena for de-fining, challenging, and refining their social identities and status in the group.

"IT SEEMS TO ME THAT THIS FRANCO KNOWS A LITTLE BIT ABOUT EVERYTHING": CONFLICT AND DISPUTES AMONG ITALIAN CHILDREN

Carlo and Paolo are building a castle with Lego-type building materi-als in the Bologna preschool. During their play they accidentally knock to the floor a castle that Alberto constructed earlier. Alberto now returns.

"What happened?" Alberto demands.

"I don't know," says Paolo.

"You don't know? It's a disaster!" says Alberto.

"It's Carlo's fault," says Paolo, blaming his friend.

"No. It's not true," denies Carlo. "It's Paolo's fault."

A boy, Stefano, who was playing nearby with some other kids comes over and says to Alberto, "It is the fault of Carlo and Paolo, understood?"

"Yes. Yes. Understood," says Alberto, nodding his head in agreement with Stefano as he begins picking up the pieces of his broken castle.

One of the first things that struck me in my observations in the Bologna preschool was the complexity of disputes and debates in the children's peer culture. In this instance, Carlo and Paolo have carelessly knocked Alberto's construction to the floor. When Alberto returns and sees his broken castle, he does not run to a teacher to complain as many of the middle-class American children I observed did in such situations. Nor does he directly accuse Paolo and Carlo. Instead he simply asks what happened. Paolo first reports that he does not know, which Alberto dismisses as nonsense; how could such a disaster happen without anyone noticing? Realizing that playing dumb is not going to work, Paolo and then Carlo deny responsibility by blaming each other. At this point a third party, Stefano, enters the dispute. Third-party entry of this type is very common in Italian children's disputes, but rarely occurs among the white middle-class American children. Stefano, who saw what happened and overheard the dispute, points out that both Paolo and Carlo were responsible and ends up with the tag often heard in Italian children's disputes, "*Hai capito?*" ("Have you understood?"), to stress his version of the event to Alberto. Alberto quickly accepts Stefano's explanation as the "true facts," which he had suspected all along.

What the kids have done here is to take this disputable event (damage to the play construction of another child) and turn it into what Italians refer to as a topic for *discussione*. In such debates, one's style or *"metodo di persuasione"* is more important than any eventual resolution.

Discussione and the Cantilena Among Italian Preschool Children

Although the Italian kids sometimes engaged in rather simple disputes over objects or the nature of play, they much more often engaged in complex debates about their knowledge, beliefs, and opinions. To get a flavor of the complex *discussione* among the Italian kids, let's consider three examples. We begin with some segments from a long debate among three boys (Dante, Mario, and Enzo, all about six years old and in their last year at the Bologna preschool) whom we discussed in Chapter 3. The boys have just finished playing a board game and are now considering a new play alternative. Dante suggests playing with Clipo (a type of building material that he is especially adept in using to build spaceships and other objects). Enzo and Mario immediately reject this suggestion and a debate ensues that first centers around Dante's expertise in building things with Clipo. Enzo agrees that Dante does indeed construct beautiful things with Clipo, but argues that this is because Dante has Clipo at home. Therefore, when he watches cartoons with spaceships he can practice building models of them. Dante rejects this interpretation of his expertise and the debate then turns to the changes in peer culture that will occur when the children leave the preschool and move on to elementary school.

"Yes," says Enzo, "but then you'll see that after a while you get bored playing with Clipo. Understood, Dante?"

"No, this is not true," denies Dante.

"Now do you want to bet that I'm right," says Enzo, "because when you're 20 and already know how to write, Clipo you won't have it, you won't have them anymore these little toys."

"I know Enzo," says Dante, "but really I with the spaceships—."

"And you'll say," interrupts Enzo, "Clipo, Clipo is silly! Because you know that if Clipo is also on television, it's not such a good thing."

In this part of the discussion, there is a shift away from talk about expertise in making things with Clipo to a more general evaluation of the activity as an important feature of peer culture. In a series of turns, Enzo skillfully dismisses the importance of Dante's expertise with Clipo by arguing that the activity is a child-like pursuit that will be left behind when the children grow older.

We see that Enzo is looking to the future and arguing that activities and objects valued in the present will pale in comparison to future alternatives. Although Enzo is overestimating how long it will take to learn how to write, he is aware that instruction in writing begins in elementary school and that the acquisition of this and other skills will lead the children to look back on activities like playing with Clipo as silly and boring. In short, Enzo is telling Dante that to remain part of this friendship group, it is necessary to look to the future and not cling to the childish activities of the present.

In another phase of the discussion, Dante makes a final attempt to hold his ground in the debate with Enzo and Mario.

"Because really, listen," pleads Dante, "I always play with it in order to build spaceships and see how they come out and to see if it is possible to make them. When I'm grown up, I'll really be able to do the work that I like because I—."

"Yes," agrees Enzo, "but first you have to practice doing that stuff. It's not that you choose for yourself a job when you don't even know how to do it."

Dante's argument is complex. First, he gains the floor with the stylized phrase "Because really, listen." Attention markers of this type are frequently used in Italian adult and children's *discussione* to both secure a turn at talk and to signal a coming disagreement to the previous claims of others. Dante goes on to note that he does indeed play with Clipo to build spaceships, but he notes that he does so "to see how they come out to see if it is possible to make them." This last phrase is striking in that Dante relates his building of spaceships in play to the possibility of the more serious activity of design engineering. In short, Dante is attempting to link the activities of the current peer culture to his perception of possible adult activities in his future. This attempted linkage is highly creative in that he is implicitly arguing that adult engineers work first from models not all that different than the ones he creates with Clipo.

Enzo quickly interrupts Dante, however, and argues that it is not possible just to choose a profession so easily. He maintains that it takes training and practice. Again the boys' perception of the timing and nature of the socialization process is striking. Enzo seems to be saying that Dante's linking of preschool peer activities to adult professions involves too big a leap of faith and that occupational socialization is much more complex than he thinks it is.

In the second example of *discussione* from the Bologna preschool, two girls (Luisa and Emilia) and two boys (Franco and Stefano) are playing with building materials. They have constructed a city, with each child adding new buildings as the play progresses. These kids have played together before, but the two boys played together often and see themselves as good friends. Although Emilia and Luisa do not have such a special friendship, Emilia has come to the defense of Luisa when the boys tease her by calling her *Genoveffa* (a teasing term meaning unattractive girl; for example, Genoveffa was the name of one of Cinderella's stepsisters). Emilia relishes taking on the boys in the *discussione*, which involves debates about the kids' knowledge of certain facts, including how the microphone I am using to tape the discussion worked. In fact, a reference to the microphone sets off the following exchange.

"Bill, ciao Bill," says Stefano into the microphone.

"But what are you saying," chides Emilia, "only ciao Bill?"

"You are saying ciao Bill," answers Stefano.

"Are you trying to make some shit with Bill?" says Franco teasingly to Emilia.

"Shut up, Franco!" says Luisa.

"Shut up, Genoveffa," counters Franco.

"But what you are saying is being recorded, you know Franco," cautions Stefano.

"Do you know that it [seems to be referring to the microphone] eats everything?," says Franco.

"Yes, it eats Genoveffa. It eats everything, eh?" says Stefano.

"Yes, anyway afterwards—," begins Emilia.

"If you don't stop it," interrupts Franco, "the record will eat you because you are talking too much."

"But excuse me," responds Emilia, "if we must talk, what can we do? Should we stay mute?"

"Ah yes," says Franco, "we stay mute. Ah—ahh—ahh, OK?"

"It seems to me," observes Emilia, "that this Franco knows a little bit about everything."

"It seems to me," says Franco, "that we need to make a cake."

The debate over the microphone emerges when Emilia berates Stefano for talking into it and saying something as simple as "Ciao Bill." Emilia's turn is highly stylized in that she subtly and effectively makes her point. She begins with a rhetorical question ("But what are you saying?") followed by a repetition of Stefano's "ciao Bill" modified by the adjective "only." Emilia, thus, first draws attention to and then mocks Stefano's behavior.

Stefano responds with a weak retort that makes little sense, a simple inversion of Emilia's claim. Franco now enters the fray using a slang expression accusing Emilia of "making some shit with" or playing up to me, which prompts Luisa to tell him to shut up. Stefano cautions Franco, reminding him that what they are saying is being recorded. This caution does not sway Franco, who makes several silly comments about the microphone eating things. He then goes on to interrupt and reprimand Emilia for talking too much.

Franco is setting himself up for a fall, because Emilia deftly dismisses his reprimand. Here, she first draws attention to her coming response with the insincere apology "But excuse me." She then goes on to demonstrate the absurdity of Franco's claim that they are talking too much by asking if they should remain mute when they have something to say. Franco responds by agreeing they should stay mute, and begins to make sounds as if he can't speak properly.

Emilia scoffs at Franco for his silliness, noting that he is the type of person who thinks he knows something about everything. She starts with the phrase "It seems to me," which links her later characterization of Franco to her as a speaker in a remarkable way. The phrase implies that she is a logical and right-minded person (it would seem to her and any reasonable person). Then there is the strategic use of the demonstrative adjective "this" to modify Franco. Used in this way, the demonstrative adjective depersonalizes Franco and sets him up for the negative characterization that follows.

Finally, the negative characterization itself is indirect in that we infer that "one who knows a little about everything" really "knows a lot about nothing." Franco himself confirms as much when he borrows Emilia's "It seems to me" but fails to do anything with it except suggest that they "make a cake."

This example demonstrates how the boys' attempts to solidify their friendship by teasing Luisa opened them up as targets for Emilia's skills in *discussione*. Emilia seized the opportunity to display her verbal talents in debate, to the relief of Luisa and the dismay of Stefano and Franco.

Our final example of *discussione* among the Bolognese kids is similar to the debate about "who has the biggest cable" among the Head Start kids. A group of kids (Sara, Franco, Giovanna, Nino, and Luigi) are sitting around a table drawing. It is *"designo libero"* ("free drawing"), so the kids can draw whatever they like. As with the Head Start kids, the discussion begins with two children having a *discussione* that they eventually draw the whole group into.

Franco draws a picture and says it is an "extraterrestrial tree." Sara waves her hand and says that "they don't exist." Franco insists that they do. A little later Franco draws what he says is a werewolf or bad wolf. Again Sara challenges Franco, saying "Wolves do not exist."

"Yes, wolves exist," says Giovanna.

"They don't exist," counters Sara, "only their bones."

"It's not true," protests Franco. "Wolves do exist!"

"Yes," agrees Luigi.

"But they do not exist," insists Sara, "only in the mountains."

At this point a boy, Paolo, who was painting at another table, comes over and, waving his paintbrush, says, "It's true. They exist!"

Sara waves Paolo away with her hand, saying "You're not in this."

Franco, now visibly upset, pokes his finger at Sara's chest and says, "You're not in this because—."

Sara pokes back and interrupts Franco, "You—."

"You say that I'm not in this." Franco interrupts right back as he pushes Sara's hand away. "Wolves exist!"

"No, it's not true," denies Sara.

Paolo, not put off by Sara, leans forward between her and Franco and says, "Not even ghosts."

"It's true," says Franco.

"The ghosts—," starts Luigi.

"Yah!" Franco interrupts. "They don't exist."

"No. No. Those no," Sara agrees.

"Yes," says Franco, now changing his mind. "Yes, they exist. Ghosts, however, exist—."

"They're in the woods," Nino interrupts.

"Eh, it's not true," says Franco. "Ghosts exist under the sea in houses—."

"In—in abandoned houses," says Paolo, finishing Franco's sentence.

"It's true," says Franco, "underwater houses."

"In the—the dark houses," chants Sara, in a sing-song cadence the kids call the "*cantilena.*" She continues in the *cantilena* and claps her hands as she says, "They stay in the dark."

"Yes, it's true," chants Paolo in agreement.

"And under the sea—it's dark," chants Franco.

"Yes, it's true," Nino chants.

"An under—they go there," chants Sara.

"No," chants Luigi, as he hits his hand with a marker, "also crabs go there."

"Submarines go there," chants Franco.

"And also sharks. And also sharks." chants Nino, as he claps his hands.

At this point, several children start talking at once and the *cantilena* ends, as does the discussion about wolves and ghosts.

This example, which contains many of the rich and dramatic elements of Italian *discussione,* begins over a dispute that could never really be settled: the possible existence of an extraterrestrial tree. When Sara denies that such a tree exists, it sets off a competition between her and Franco. When Franco later draws a werewolf, Sara insists that they also do not exist. Now, Giovanna takes Franco's side, claiming wolves do exist. Sara then gives in a bit, saying that only their bones exist. With this argument, Sara implies that werewolves might have existed in the past.

At this point, several other children join the discussion, including Paolo, who was not drawing at the table with the other kids. Paolo, who was painting on the other side of the room, comes over with paintbrush in hand. He stands between Sara and Franco and argues that wolves do indeed exist. Sara now does something unheard of in Italian *discussione*—she tries to exclude Paolo. She waves him off, telling him in so many words that this debate "is not his business." Franco imme-diately challenges her action by throwing the same phrase back at her and poking his finger at her chest. In this way, Franco is challenging Sara's violation of a basic rule in Italian *discussione*: Everyone has a right to be part of any discussion. Sara pushes Franco's hand away and tries to rebut his claim, but Franco says, in essence, "Who are you to say I'm [or anyone is] not in this." He then goes on to argue again that wolves exist.

Paolo, now a full participant, adds a new element: that ghosts do not exist. Franco first agrees with this claim, then seems to change his mind, perhaps because Sara also agrees that they do not exist. At this point there is general discussion about ghosts and where they can be found (in the woods, in abandoned houses, and finally in abandoned houses under the sea). It is not clear where the kids came up with these ideas about ghosts—perhaps from stories read to them or from watch-ing cartoons or movies. What is clear is how the kids are able to work together to develop the discussion, with Paolo even finishing Franco's sentence at one point.

After the talk about ghosts living in abandoned houses under the sea, Sara takes a turn at talk in the *cantilena*. The *cantilena* is a tonal device or sing song chant that the kids often used in *discussione*. Sara's use of the *cantilena* is especially impressive because it involves three separate phrases all produced in the falling and rising pitch and each containing different elements related to previous discussion (ghosts, dark houses, ghosts underwater). It is hard to appreciate the phonetic aspects of the *cantilena* without hearing it. American children, at times, present a similar verbal routine that goes something like: "My Dad's bigger than your Dad!" However, such verbal play seldom goes be-yond a few exchanges. The *cantilena* is more complex and has no typi-cal and predictable verbal content. Instead, the children have to fit the content of ongoing *discussione* into the structural demands of the *cantilena*.

For example, to preserve the sing-song pitch, it is necessary to produce a phrase with at least four syllables, and one has to think of something to say of this length quickly that fits the ongoing discussion. Long turns with new information are especially difficult to produce, and challenging turns at talk in the *cantilena*, like Sara's, are appreciated by her peers. On the other hand, minimal participation keeps the *cantilena* going and is also valued. For example, Paolo's "*Eh, è vero*" ("Yes, it's true") works in the *cantilena* because he adds the "*Eh*" (or "*Sì*" or "*No*") before the "*è vero*" providing enough syllables to work with.

After Paolo's response, Franco, Nino, and Luigi all use the *cantilena* to contribute to the discussion. They signal either agreement or disagreement, refine previously mentioned information (it's under the sea), or add new information (other things that are underwater like crabs, submarines, and sharks).

What is most important about the *cantilena* in *discussione* is that it is a consciously shared element of the peer culture. That is, the kids are aware of their use of the *cantilena* and they use it to dramatize and enliven *discussione*. In fact, the kids' frequent chanting in the *cantilena* sometimes irritates parents and teachers, who discourage its use with the command "*non far la cantilena!*" ("Don't do the *cantilena!*"). Interestingly, in family role-play in the Bologna preschool, the kids pretending to be mothers, fathers, older siblings, or teachers often use this command when their pretend charges produce the *cantilena* in pretend discussions and quarrels. In this way, the children take the adults' disapproving reactions to their creation of the *cantilena* and embed them into their shared peer culture in play. We again see how peer play routines can be used to challenge adult authority.

Negotiations and Peace Among Italian Preschool Children

Earlier, I noted that some researchers have argued that children's conflicts and disputes are rarely negotiated and settled by the children themselves because adults are so quick to intervene. Mostly, adult interventions impose settlements on kids because adults have more power. On many occasions, kids might not be happy with the outcome, but they usually accept it and move on.

However, I found that the teachers in the American Head Start program and in the Italian preschools were less likely than teachers in the middle-class schools in Bloomington and Berkeley to intervene in children's disputes at the first sign of conflict. Therefore, the children's disputes were longer, more complex, and often developed from spats between two or three kids to group debates. In the examples of group debates that we discussed in this chapter, the initial dispute was not always clearly settled. However, any serious conflict dissipated and the kids went on to more general discussion, where they tied their contributions to personal experiences and honed their skills in debate and argument.

In the Modena, Italy, preschool there were many instances of long group debates like the ones described. There were also more serious disputes where the kids involved and those who overheard disputes worked hard to establish agreement and peace. Remember that the Modenese kids had been together in the same group with the same teacher for three years. They worked hard to preserve their strong group identity, as we saw in the incident described in Chapter 3 in which the kids spent considerable time and energy to bring two friends who got into a serious tiff back together.

Although the Modenese children valued discussion and debate, they often worked hard to negotiate agreements. Here's a nice example where I am brought into the negotiations, but strictly because of my size and not my intelligence or negotiation skills.

The Hair Debate

Marina and Sandra are playing with dolls and Sandra insists that one of the dolls (an infant) with little hair must be a boy because it has short hair. Marina disagrees. She says that babies, both boys and girls, often have short hair. But Sandra disputes this claim, again saying that only boy babies have short hair. Some of the children playing nearby join the discussion. Some side with Marina, some with Sandra. Marina then points to the shelf where the children's personal books or portfolios (which document the children's lives and time in the preschool) are stored. She asks me to reach up to get her book down because she can't reach it. I do so and Marina says, "Grazie Bill," as I hand her the book. She then turns to a page with a picture of her when she was about a year old. (Each kid's book has a baby picture.) Sandra and several other kids gather around to look at the picture. We all see that Marina had little hair in her baby picture. "See," Marina says to Sandra. "This is me and I had short hair then." Sandra now says, "*Hai ragione*" ("You're right"), and the issue is settled to everyone's satisfaction.

Marina's use of me in this episode is interesting because she relies only on my size, which enables me to reach up and get the book. She does not ask for my support of her position and does not assume that I know any more than the kids about the disputed topic. She might have refrained from asking for my support in the group because she knows that I have not been in the school for the three years that she has, and also because (as discussed) the Italian kids see me as a somewhat incompetent adult. However, it is also the case that these kids often feel that they can handle these types of disputes on their own and do not want to turn to adults for help.

The example also shows how the kids take an element of their collective experience in the school culture—the existence of the personal books that they have created about their experiences over the three years that they have been there—and use it to address a dispute in the peer culture. In doing so, the kids feel empowered to solve their own problems, without adult intervention.

Here's another brief example, which displays many of these same themes.

Pace, Pace, Carote, Patate

Several children are sitting around a table with workbooks, which the teachers encourage the children to work on at their own pace to develop their literacy skills. The books contain various tasks, including drawing pictures next to words or short texts, linking scrambled texts and pictures, filling in missing letters of words, and so on. Luciano makes a negative remark about the quality of Viviana's drawing while she works in her workbook. Viviana becomes upset and the dispute escalates, with Viviana telling Luciano to mind his own business and commenting that his drawings are not perfect. The two go back and forth about this and several other children try to appease them. However, none of the kids calls the teachers to help. The teachers overhear the dispute, but do not intervene. At one point, having grown weary of the arguing, another girl at the table, Michela, says, *"Adesso basta. Pace!"* ("Now enough. Peace!"). Viviana and Luciano agree to end their argument and a little later are laughing and joking. They even produce a rhyme: *"Pace, pace, carote, patate!"* ("Peace, peace, carrots, potatoes!").

Here again, we see the active involvement of kids to settle their own disputes. Interestingly, a girl not involved in the dispute grows weary of the bickering and demands peace. The bickering kids, Luciano and Viviana, agree and then they, who were so upset with each other earlier, go on to mark their establishment of peace *through the creation of a literacy activity*. They create a poem that is an impressive play on words in Italian—a poem that is funny as well as creative.

Here's a final (somewhat longer) example that captures the competition that existed between the two groups of five-year-olds in the Modena preschool and the kids' determination to make peace in very challenging circumstances.

La Guerra Dell'Erba (The Grass War)

The outside yard of the preschool has been freshly mowed and cut grass is lying all around. Some of the girls (Elisa, Carlotta, and Michela) begin gathering the grass and take it to an area under the climbing structure, where they make a bed. At one point, Michela and then others lie down on the bed and say: *"Che morbido!"* ("How soft it is!"). Several other girls enter the play, but Elisa, Carlotta, and Michela control the activity. The new recruits are allowed to bring grass, but are not allowed to place it on the bed.

Later, Carlotta returns to say that one of the boys from the other group of five-year-olds at the school hit her while she was gathering grass. The other girls decide to go get the boy. They march over, carrying grass, come up behind the boy, and pummel him with the grass. The girls then run back to the climbing structure and celebrate their revenge—especially Carlotta, who is all smiles. Eventually, the boy gets a few of his friends and they come by and throw grass at the girls. The girls chase after the boys, who are outnumbered, and take the worst of it in another exchange of grass throwing.

The grass war now escalates with girls and boys on both sides becoming involved. In fact, all but a few of the group that I am observing are now in the grass war. The war continues for some time until Marina suggests to the children in our group that they make peace. Marina, with several children behind her, marches up to the boy who hit Carlotta and offers her hand in peace. The boy responds by throwing grass in Marina's face. Marina returns to the group, and Carlotta says: "They don't want peace!" But Marina says she will try again. The second time she offers her hand, the boy throws grass again, but over the objections of another boy who is in his group. Marina stands her ground after being hit with the grass. The second boy pulls his friend aside and suggests that they make peace. The first boy is against the proposal, but eventually agrees and the two boys shake hands with Marina. Marina then returns to our group and declares: "We now have peace!" The two groups meet for a round of handshakes. I also exchange handshakes with the kids from the other group, who identify me as part of the opposing group.

In this episode the kids from 5b, whom I was studying, appropriate objects (the freshly cut grass) from the adult world and use them to create an innovative pretend play routine, a creative activity that gives the children a shared sense of control over their social environment. The inter-group conflict between the two groups is both related to, and further develops, the strong solidarity within the 5b group. Later, the peace negotiation, symbolically marked by handshakes, demonstrates the children's awareness of a sense of community in the school. My inclusion in the handshakes confirms my place in this community.

CONFLICT, ADULT CULTURE, AND CHILDREN'S PEER CULTURE

Conflict is a central feature of kids' peer culture. However, children's attitudes toward and engagement in conflict and debate are very much part of their experience in their local school cultures and in the wider community and society of which the kids are members.

Conflict among the white middle-class American children was often seen as negative and threatening by the teachers. This reaction to conflict and disputes was shared by most of the parents. In teacher-parent meetings that I attended in the American middle-class schools I studied, parents stressed the importance of children "talking over" problems, rather than resorting to physical aggression or even verbal disputes. They wanted their children to get along and play nicely, and therefore, almost any form of conflict was seen as threatening. As a result, the kids were somewhat thin-skinned and conflicts (especially those related to friendship, as we saw in Chapter 3) were often fraught with emotion. At the same time, these emotionally charged events led kids to reflect on the nature of their peer relations and their more general position as individuals in the peer culture.

Although the American Head Start children's oppositional style in peer play might seem aggressive to most white middle-class Americans, it was seldom interpreted that way by African-American kids, teachers, or parents. As the anthropologist Roger Abrahams has argued, among African-Americans, opposition and conflict tend to be viewed as consistent antagonisms "that cannot be eliminated and in fact may be used to effect a larger sense of cultural affirmation of community through a dramatization of opposing forces." In other words, oppositions and challenges among the Head Start kids were dramatic exchanges that were reacted to in kind, and the overall tenor of exchanges was playful banter. This verbal dueling sent two messages: (1) that a particular child could hold his or her own ground and not be easily intimidated and (2) that participation in oppositional talk signified allegiance to the values and concerns of the peer culture.

Having been exposed to and included in *discussione* by parents, teachers, other adults in the community, and siblings, the Italian preschool children generated, valued, and refined the dramatic verbal style in their own peer culture. In the course of such *discussioni*, the children shared a sense of collective identity, formed friendship alliances, and developed and displayed personal skills in discussion and debate.

Overall, our examination of conflict in kids' cultures helps us to see the importance of diversity in the lives of children and adults. We can develop a better appreciation of the complexity of kids' cultures by remembering that they arise out of the highly diverse and complex adult cultures and societies in which they are embedded. However, to appreciate fully the complex features of the kids' cultures I described in this book, it is necessary to take children and their childhoods seriously. We must resist the typical tendency to "look down" on our children from above and to overprotect, undervalue, and even discriminate against them. In the concluding chapter of this book, I challenge all adults to shake the many misconceptions of our children and to appreciate, embrace, and become more actively involved in their worlds.

8

"Appreciating Childhood"

. . . .

Suggestions for Supporting and Sharing in Kids' Culture

In my research with young children the end of a particular study is always bittersweet. The kids and I reflect fondly on our time together and we know that, in most cases, we might not see each other again. Over my many years of ethnographic work, I have been able to keep in touch with a number of kids as they grew up and we are good friends to the present day.

My most recent study in Modena, Italy, was special in this regard because I joined a group of kids and stayed with them during one of the most important transitions in their lives: from preschool to elementary school. After we had been in the elementary school for a few weeks, one girl, Stefania, whom I had first studied in preschool asked me in class, "Bill, will you stay with us all the way to high school?" The teachers and I laughed at Stefania's question because it was so cute and somewhat surprising. But for Stefania the question was a serious one. She had made a good friend in me, and I was providing her some security, which she hoped could continue throughout her childhood.

In fact, I continued to study and be with the group of Modenese children (which increased from 21 to nearly 80 after we got to elementary school) as they advanced through elementary school and entered middle school. When I returned to the school in May 2001, the children were fifth graders and nearing the end of their time in the elementary school. It was a joy to see and be with the kids as they visited the middle school they were to attend and as they served as hosts for the preschoolers who came to visit them and their teachers (the fifth grade teachers recycle and begin instructing a new group of preschoolers when they enter first grade).

The group of kids I worked with had studied English in preschool and took up their English studies again in third grade and many had become quite proficient, especially in reading and writing English. As a gift, the kids and teachers gave me a book entitled *Per Un Amico* (*For a Friend*) that contained poems with spaces to write appointments, thoughts, and memories. The children filled these spaces with messages to me in English in which they expressed their feelings and said goodbye because many of us would not see each other again. The messages the kids wrote to me were moving declarations of friendship and celebrations of our time together.

Andrea wrote his message next to a picture of a kite:

DEAR BILL,
    I HOPE THIS KITE WILL
    FLY FOR EVER LIKE OUR
    FRIENDSHIP.
                 LOVE
                 ANDREA

I have been very fortunate as an adult to share in the lives of many kids *as they lived their childhoods*. This does not mean that I have been able to have a second childhood. Even though I participate in the kids' peer cultures to some degree, I am still an adult. What I have acquired is insight, understanding, and appreciation of childhood and kids' peer cultures. I firmly believe that all adults who take kids seriously and who are open to learning from them can develop such an appreciation. They can also benefit as human beings from the experience.

Societies and cultures that are structured so that the daily lives of kids and adults are highly integrated, that have policies and customs that support and celebrate childhood, that encourage a positive attitude that all children are part of one cultural family, and that ensure an equitable distribution of resources across age groups—these are the societies and cultures that will prosper and lead us in this new century. But why do so many societies, including the United States, fall short of these goals? And how can they do better?

YOUR CHILD, MY CHILD, OUR CHILDREN

In the United States we have a very mixed attitude toward our children. On the one hand, we say that our children are our future and that we must protect and invest in them. On the other, we see children primarily as the individual responsibility of their parents or parent. Therefore, we tolerate the fact that many children live in poverty and lack basic social services such as quality child care, early education, and adequate health care. And in some cases we even actively discriminate against children.

Let's begin with the discrimination. Sometimes, because we do not see children as fully developed humans, discrimination is subtle and unwitting. Other times it is planned, obvious, and glaring. Regarding the former, consider the tendency to label children for the failings of adults. On the cover of a recent book on overindulgent parents, *Too Much of a Good Thing: Raising Children of Character in an Indulgent Age*, by Daniel J. Kindlon, we see a four- or five-year-old girl who pulls her mouth open with her fingers to make an ugly face at potential readers—the image of a spoiled brat. I have to ask: If the book is about bad parenting, why depict a child in a negative way and not the parents?

Such negative stereotyping of children is quite common in American advertising generally. Every year, various companies selling back-to-school supplies run commercials celebrating what they call "the most wonderful time of the year," that time being when parents can get rid of their kids because they have to return to school from summer vacation. In one commercial, two kids get into the family minivan happily chanting: "We're going to the water park!" But then once locked in the van, their parents gleefully tell them: "You're going back to school!" When the kids try to get out, the door handles come off in their hands. Even worse, in another ad a preadolescent girl is talking on the telephone in a stereotypical way, "Like Wanda said Jake doesn't like Brittany—," when suddenly her parents come into the room and a cage drops from the ceiling to capture her. The parents then wheel her out, still in the cage, to a waiting school bus. The girl will have to talk with her friends at school more often now rather than tying up the family phone and annoying her parents.

Adults often see such commercials as "good fun" and not offensive. However, what do kids think of them? To kids I bet they're not so funny. We can easily envisage the negative response from any other group in society if it were depicted in such a way. Imagine the reaction to a commercial showing a family hauling off grandma in a cage to a retirement home!

Such ads are not surprising in a society that persistently uses the word "childish" or the phrase "acting like a child" instead of the more correct term "immature." Such usage leads to bizarre statements that label adults who have adulterous affairs as "acting like kids" rather than labeling this behavior with the obviously appropriate word "adultery." Finally, there is the offensive, and fortunately recently less commonly used, label "illegitimate child." *There is no such thing as an illegitimate child because children have no control over their parentage.* If a negative label is to be used in this case, it should label the behavior of the adults involved, not their children.

In the United States we do not stop at the negative stereotyping of children, but permit outright discrimination. There are many places (apartment complexes and gated communities) in the United States where children are not allowed to live. In these communities "adults only" rules are often enforced with no exceptions, as was the case in Arizona where a preadolescent was evicted from the home of his grandparents where he had been living temporarily while his mother was in a drug rehabilitation center.

To get a feeling for how blatant such discrimination can be, consider the following pages, shown in Figure 5, from a brochure sent to me advertising an "adult only" cruise.

On the next page we see a picture of a red-haired, freckly faced boy making a face by sticking out his tongue and wiggling his hands with his thumbs in his ears. The caption to the picture reads: "here's someone you won't run into." Under the picture we're told: "You'll enjoy a 'grown-up' atmosphere on our new . . . ships. So, with no one under 18 years of age on board, your Cruise Tour Vacation will be relaxing and kid-free." On the page after next, some examples of this "grown-up" alternative cruise are described with pictures and text. At the bottom is another caption that reads: "Smoke-Free, Kid-free Environment Throughout!"

Kids are actually equated with smoke! Again replace the word "Kid" with any other group in society: Man, Woman, Elderly, African-American, Native American, and so on. The resulting uproar would put the company out of business—and rightfully so. But with kids hardly anyone seems to notice or view the ad as an example of blatant stereotyping and discrimination.

One reason discrimination against children often goes unnoticed is that children are not seen as full members of society—as citizens with basic rights and privileges. Instead, in the United States children are seen as extensions of their parent or parents, who are ultimately responsible for them. The notion that "it takes a village to raise a child,"

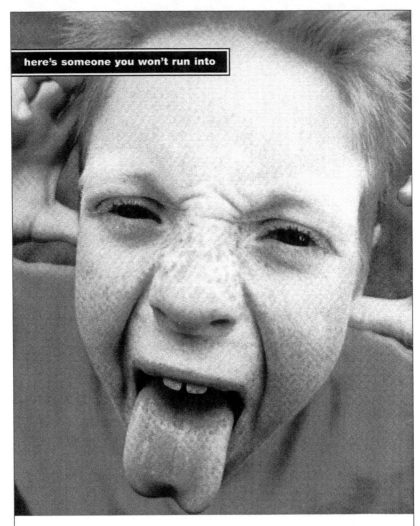

here's someone you won't run into

You'll enjoy a "grown–up" atmosphere on our new 684–passenger R-Class ships. So, with no one under 18 years of age on board, your CruiseTour Vacation will be relaxing and kid-free.

FIGURE 5 Kid-free cruise.

The ∧ alternative.
grown-up

ENTERTAINMENT From the "Great White Way " of Broadway, to the red hot jazz of New Orleans, there are musical tunes to suit every taste. Try your luck as a game show contestant, hit the Karaoke stage, or experience the flavor of local performers who come aboard to entertain you.

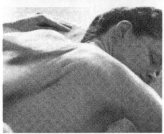

RELAXATION Try a facial in our full-service spa and a relaxing massage from our European-trained therapists. Experience an invigorating Aromatherapy session to stimulate your senses or a Thalassotherapy whirlpool massage where you will feel the soothing effects of warmed seawater.

HEALTH AND FITNESS Our shipboard wellness facilities include a fully equipped and professionally staffed fitness center, a spa offering holistic treatments, and healthy menu choices. Plus, our exclusive relationship with Johns Hopkins Health Systems provides access to quality medical care.

Smoke–Free, Kid–free Environment Throughout!

although a popular one among some progressive politicians and child advocacy groups, does not generally mesh well with mainstream values of the competitive, democratic, socioeconomic system of the United States. Americans generally believe in hard work and self-sufficiency and mistrust big government social programs, especially for families with children. There is universal support for kindergarten through 12th grade education, administered at the local government level. As a result, school systems vary widely in terms of resources and per-pupil expenditures. There are also means-tested (you have to be very poor to qualify) social welfare and educational programs to provide a "safety net" for the poorest American children.

However, these programs pale in contrast to universal social welfare programs for older Americans, for example, Social Security, Medicare, and Medicaid. In fact, as former Secretary of Commerce Peter Peterson has pointed out, the United States federal budget dispenses nearly 10 times as much in benefits to each adult aged 65 years and older as it does to each child aged 18 and younger. These programs for older Americans are good ones and have been successful. They are one of the reasons older Americans are living longer and enjoying a high quality of life. However, their continued success depends on investing more in children, whose production as adults will keep these programs for older Americans solvent, especially as the baby boomers reach retirement age.

In short, in the United States we have an inequitable distribution of social welfare spending with much more invested in older Americans than in children. The safety net for children is minimal. Programs like Aid to Dependent Children and food stamps have barely kept up with inflation over the last 20 years, and in most states families can hardly make ends meet, given the level of benefits these programs provide. Resources for Head Start have increased in recent years, but in most states only four- to five-year-olds are covered and many states provide only half-day programs.

Furthermore, many poor children of working families do not qualify for programs like Aid to Dependent Children, food stamps, Head Start, and health care through Medicaid. In the United States more than 16 percent of children live in poverty and nearly 12 percent are without health care insurance. High-quality child care and early education are a major problem for the working poor and even middle-class families. Until 1993 the United States was the only industrialized country in the world, except South Africa, with no formal policy for maternity or family leave. The United States Family and Medical Leave Act mandates employers of 50 or more employees to provide up to 12 weeks of *unpaid*, job-protected leave to employees for certain family and medical reasons. Reasons for the leave include the birth of the employee's child, or its placement as an adopted or foster child; the care of the employee's spouse, son, daughter, or parent who has a serious health condition; or a serious health condition that makes the employee unable to perform the job. Although the Family and Medical Leave Act was clearly a step in the right direction, especially because of its guarantee of job protection, the fact that the leave is unpaid is a major shortcoming. Most working-class and low-income families (especially single-parent families) cannot afford to take more than a week or so off from work because of the lost income. Further, once the parent returns to work after some period of family leave, many working-class families must rely on substandard child care, which, though it might be poor quality, can consume 25 to 30 percent of family income.

The quality of life of families with children is much higher in other modern societies than in the United States. Although the percentage of children living in poverty in a few industrialized countries is nearly as high as that in the United States (for example, Italy, which has a wonderful early childhood education system, has a child poverty rate of 14 percent), most countries have poverty rates half that of the United States, and several countries have rates less than 3 percent (Denmark, France, Norway, Austria, and Luxembourg) and three have rates less than 2 percent (Sweden, Belgium, and Finland). All Western European countries provide universal health care for children and are far ahead of the United States in terms of family leave. All Western European countries also provide maternity leave at 50 to 100 percent of earnings—from six weeks to one year. Most countries provide additional family leave (normally during the child's first year) at some percentage of earnings or at a flat fee.

In Western Europe, maternity and parental leave policies address infants' needs in the first year; custodial care programs are available for the one- to three-year-old children of working mothers; and early education programs are normally provided for *nearly all* three- to six-year-olds. The costs and availability of programs for toddlers are similar to what we find in the United States. But in Western Europe, most of these programs are government subsidized to reduce costs for parents and increase quality. In fact, in some countries, programs for toddlers have evolved from being mainly custodial to being more educational in approach.

The biggest difference between the United States and Western Europe is the extensiveness and overall quality of early education programs for three- to six-year-olds. Almost all European countries offer quality programs at a low cost. Although early education teachers have generally less training and are normally paid somewhat less than elementary and secondary school teachers in these countries, their incomes are much higher on average than early education teachers in the United States. The Western European preschool programs have much lower rates of teacher turnover than those in the United States, and there is more parental involvement and respect for teachers there. In countries like France and Italy, more than 95 percent of all three- to six-year-olds attend government-supported early education programs, and parents pay very low fees—primarily to cover the cost of meals. Programs in these two countries are seen as exemplars of the best early education in the world. They are based on carefully developed early childhood curricula that stress social and language skills and bridge the child's transition from the family to the community and formal schooling.

IMPROVING THE QUALITY OF OUR KIDS' LIVES AND APPRECIATING THEIR PEER CULTURES

In this book I have provided an insider's view of the complexity of young children's peer cultures based on my research in several preschools in Italy and the United States over nearly 30 years. In all the preschools I studied the kids created vibrant peer cultures that reflected the joy, wonder, and complexity of their childhoods. In the American preschools the complexity of the kids' culture was, by and large, hidden from most adults beyond the teachers and a few parents. In the private, not-for-profit preschools I studied in Berkeley, California, and Bloomington, Indiana, the overall quality and resources of the schools were high and there was some parental involvement. Not-for-profit private preschools often provide high-quality teaching, staffs, and curricula as well as resources, because tuition revenue is used only to pay the program's staff. For-profit private preschools, on the other hand, are normally of lower quality because teacher pay and investment in resources are often kept to a minimum to increase profits. Parental involvement in the private preschools I studied normally occurred outside the classrooms and was primarily administrative, especially by parents who served on the preschools' boards of directors. There were, however, a few special events throughout a given school year, like family dinners and musical and other talent performances by the children for their families. There were also a few field trips into the local communities, like visits to zoos or museums. These events helped to bring parents and kids together and also involved the kids in the wider community to some degree.

The Head Start programs I studied lacked the resources that were typical in the American private preschools or the Italian government preschools I studied. Although federal government support of Head Start has increased since the time of my study, the programs are still limited in terms of the number of children covered and the range of program activities. The programs were in general of good quality but were limited to a half day, some portion of which was spent on structured language and cognitive tasks based on a compensatory orientation. That is, this part of the curriculum assumed that poor children lacked cognitive and language skills and needed such drills to catch up or have a "head start" for kindergarten. I found, on the other hand, that the drills (while teaching children to raise their hands and take turns in lessons in ways comparable to public school kindergarten and first grade practice) often confused the kids and led them to believe that there was only one right answer for any question. As a result, I found that many of the children I followed into kindergarten and first grade struggled somewhat as they became perfectionists about their work. They were hesitant to offer answers in structured learning tasks and often did not finish their work on time.

Despite the half-day limitation and problems with the compensatory nature of some aspects of the curriculum, the Head Start programs were highly valuable with respect to the kids' social and emotional development. The kids developed a strong peer culture, which was supported by a group of caring adults in the Head Start centers. In fact, the Head Start centers are best described as small communities that emphasize collective values and provide something approaching an "extended family."

In a typical day at the centers, the average child came into contact with a wide range of adults: teachers and teaching assistants, visiting parents, bus drivers, administrative staff, social workers, speech therapists, custodians, and cooks. Although the kids spent the overwhelming majority of their time at the centers with the teachers and assistants in their particular classrooms, they knew all the adults' last names and frequently exchanged greetings and playful talk. These adults also knew the children well. In fact, some adults had particular favorites among the kids and they talked and joked with them in the hallways and when they entered the classrooms.

This strong support from caring adults was very important for the Head Start children, many of whom came from single-parent families and lived in poor and dangerous neighborhoods with limited opportunities for positive interactions with adults or other kids. Furthermore, the Head Start programs encouraged parental involvement (several parents worked in the centers regularly) and the teachers visited and supported parents in their homes.

The Head Start programs also involved the children in the wider adult communities of their cities. Because the children were bussed to the Head Start center, the buses and drivers were available to take the kids and their teachers on numerous field trips over the course of the year. Many parents also participated in these events. The kids visited parks, museums, zoos, department stores for Christmas and Easter displays and activities, and workplaces like post offices and fire stations. The kids were always treated warmly by the adults they met in these various community contexts. And quite simply, the kids loved these field trips. While such activities might be taken for granted by many middle- and upper-class kids, they were special to the Head Start children. The kids talked about and anticipated the trips days in advance and often incorporated features of the activities in their peer play for many days after the events. In this way, these activities made the kids more visible in their community and also developed support and appreciation for and contributed to the kids' childhoods and peer cultures.

The Italian preschools I studied in Bologna and Modena promoted kids' construction of vibrant and complex peer cultures, which both enriched their childhoods and contributed to their development of social, language, and cognitive skills. The programs have a long history in northern Italy dating back to the 1960s and high-quality early education is the norm throughout Italy today.

Most of the legislation related to child care instituted in the late 1960s and early 1970s in Italy was the result of intense periods of social and political struggle that followed the Italian economic boom of the 1950s and 1960s. Much of the collective and highly public political mobilization of this period was directly tied to the mass migration of Italians from rural areas throughout the country to major cities, primarily in the north. This type of collective action had a long history in certain regions of the north, most especially Emilia-Romagna, the region where Bologna and Modena and the schools I studied are located. As a result, child care and early education issues were caught up in labor militancy, youth movements, the women's movement, and other urban protests.

The general orientation of early childhood education in Italy reflects the collective and communal movements from which it was born. The preschool is seen as a place of life for children. Activities such as playing, eating, debating, and working together are considered just as important as those that focus on individual cognitive or intellectual development. This communal activity is evident in the organizational structure of preschools as well as in the wide range of social, verbal, and artistic projects making up the curriculum that stress the relationship of the preschool with the family, community, and children's peer culture.

Regarding the structure of the preschools, in Bologna a mixed group of 35 children with 5 teachers attended all-day programs from September until July. Each of the three years I observed, a group of five-year-olds moved on to elementary school and a new group of three-year-olds entered the preschool. In Modena, I studied a group of five-year-olds who had been together for three years with the same teachers. This structure of keeping children together with the same teachers over the three years of preschool builds strong communal bonds among the children and between the children and parents. It is also important in parental participation in the school programs because parents get to know each other well and develop strong relationships with the teachers. In this way, the preschool is something more than an educational institution in Italy, it is often a social and community organization for families with young children.

The elements of the curriculum I found most striking in the Bologna and Modena preschools were long-term projects that involved observations, discussion, action, and reconstruction. One project in Bologna involved planning for, making, and reconstructing visits to the homes of the older kids during the spring of their final year at the school. In my first year in the preschool I was introduced to this project in an indirect way when a boy, Felice, told me: "Bill, you're coming to my house." I was not sure how to respond to this and just nodded and said, "That's good." I assumed that perhaps Felice's parents were going to invite me for a visit.

However, a few days later in a group meeting, the teachers told us about the family visits. Then each of the older kids talked about their families and the preparations they were making for our visits. All of this sounded really exciting to me and to the three-year-old kids who had not been on one of these visits before.

On the important day we walked as a group to the home of one of the older kids. I especially remember the walk to Felice's house. His home was very near the school and located in a residential and shopping area near my apartment. Thus, I knew many of the merchants with whom we stopped and chatted with along the way. The storekeepers knew about these annual outings and looked forward to the opportunity to talk with and admire the kids. We also talked with many shoppers (both men and women) as we reached the first busy street on our way to Felice's house. In fact, these conversations seemed to delay our progress from my perspective and I wondered if we would ever get to Felice's house! However, the teachers and kids were unconcerned and enjoyed the conversation and attention of the adults in the local community.

We eventually continued our journey and left the busy thoroughfare, walking down a small side street that came to an end in front of the large apartment building where Felice's family lived. As we approached the front door of Felice's building, several children ran up and took turns pressing the bell. Antonia, my partner on the walk, tugged on my arm to hurry up to the door. When Antonia finished ringing the bell, I, swept up in the moment, also gave it a long ring. Everyone laughed and one of the teachers said, "That Bill, always one of the kids. Enough. Let's go in."

Felice and his younger brother, Marco, peered down over the railing at us as we climbed the four flights of stairs to their apartment. The smile on Felice's face was unforgettable. When we arrived, we were greeted by Felice's parents and three of his grandparents (one of his grandfathers was no longer alive), all present for the big day. Felice's parents escorted the teachers into the kitchen while I was pulled off to Felice's room with the other kids. We inspected all of Felice's toys, which included an impressive collection of "*I Puffi*" (small replicas of cartoon characters—Smurfs in the United States—that were popular among the kids at that time). Eventually, we all went off to the kitchen where Felice's mother served a wide variety of scrumptious snacks. Before we left, Felice's father presented me with homemade wine and salami. That evening after I summarized the event in my notes, I reflected on my strong emotional reactions to the event and I wrote: "It was a good day!"

For several days after a home visit, the teachers and kids first verbally and then artistically reconstructed the experience. The artwork contained a series of pictures that visually captured the major phases of the event with each child contributing in some way to each picture. The detail of the pictures was striking. In a depiction of our walk to Felice's home, for example, some children drew the cars on the street,

FIGURE 6 The visit to Felice's house.

others drew individual members of the group (teachers, kids, and me), while still others drew shops and their classmates designed clothes to put in the shop windows. These pictures were then prominently displayed in the school (along with artwork related to other projects) until the end of the year when they were taken home by the older children to keep as mementos. Figure 6 is a picture of the large mural with the kids' individual artwork that depicts our walk to Felice's house.

　　In this project, the kids think about, discuss, and artistically reconstruct their relations with the school, family, community, and each other. They collectively reaffirm the emotional security of these bonds while reflecting on how the nature of these attachments changes as they grow older. In the process, the kids gain insight into their changing position in the school, peer, and wider adult culture.

In addition to long-term projects, family involvement was high in the preschools in both Bologna and Modena. In both preschools, there were "end of the year" parties where children gave singing and dancing performances to large groups of parents and grandparents. In Modena, there was both a party for the whole preschool and a special party just for my group, organized by the parents and teachers. At this party the kids performed dancing and singing routines they had practiced for many weeks. The parents and some grandparents also engaged in certain of the dances and participated in games with the kids as well as preparing a large meal. The parents presented the teachers with expensive gifts and even provided a gift (a beautiful beach towel) for me as a member of the group.

My experience with the kids in Modena was especially meaningful for me because I went on with the group to first grade and continued to keep in touch with the four first-grade classes throughout the five years of elementary school. As a result, I was able to observe and experience many of the events and activities in Modena in which a strong civic society was constructed around these children and their families. By civic society I mean a collective celebration of civic engagement, people's connections with and participation in the life of their communities. In the United States, examples of such civic society might be neighborhood block parties, bowling leagues, union picnics, and ethnic street festivals. Although such events still exist, many argue that they are fading from the lives of many Americans and their children.

Civic engagement was strong in Modena and often involved and even centered on children. In fact, preschools and elementary schools were often the site and kids the focus of many civic activities. I was introduced to this type of civic engagement in the celebration of *Carnivale* (our Mardi Gras) during my first month in the Modena preschool. First, in the school there were two days of celebration with the three- to five-year-old kids and their teachers in costume, dancing, singing, eating candy and pastries, and generally having a grand time. Then there was also a general celebration for all Modena's children, including my eight-year-old daughter and all the kids I studied, in the main square of the city. Here, kids in a wide variety of colorful costumes gathered with parents and grandparents to run, play, throw streamers, listen to music, and buy candy and other treats from street vendors.

This was one of several events that occurred both in the school and at the more general city or community level. Another was a concert of traditional children's songs performed by five-year-olds from all the preschools for all the people of the city. This performance was preceded by months of preparation and practice by the children under the direction of the music teachers in each preschool. The children had one rehearsal with all the kids and music teachers before the big performance, which was a spectacular and highly successful civic event with many proud parents and grandparents in the large audience.

Just as important, all the preparation, practice, and pride in the performance made the singing of the songs a key element in the kids' peer culture during the last months in the preschool. They often sang or hummed the songs during work on projects and in free play. I especially remember their singing of the songs at another civic event in the preschool, the *"festa di nonni"* ("party for grandparents").

Almost all the grandparents of the kids who lived in Modena attended the *festa di nonni*. There were many activities in which the kids and grandparents worked together. Some grandmothers, with girls and boys, sewed new outfits for Barbies and other dolls, while other grandmothers went up to the kitchen with a group of kids to make desserts to have after the big meal that the school's cooks were preparing. Some grandfathers worked outside in the garden with one group of kids, while another made kites. We later took the kites out into the yard, and the kids took turns flying them around. My job was fetching the wayward kites out of the trees without damaging them.

What I remember most, however, happened right before lunch. The kids sang several of the songs they had earlier practiced so often and performed for the citywide concert. I had heard these songs over and over. I now knew them by heart. As the kids sang the first two songs, I sang along with them, softly mouthing the words. In the middle of the third song, the kids, who were sitting in small chairs, laid their arms over each other's shoulders and began to sway with the music. Their faces were beaming. I looked at the grandparents. They were all misty-eyed. So was I.

"WE'RE FRIENDS, RIGHT?": A PLEA FOR COMMUNITY

The groups of kids I have studied in the United States and whose peer cultures I documented represent some of the most economically advantaged and disadvantaged in our society. They are, to a large degree, representative of these economic groups. The Head Start program gives children strong emotional support, the opportunity to develop a strong peer culture, and a certain amount of preparation for elementary school. However, these programs are only modestly funded compared to support for preschools in other countries and they lack basic resources that would expand coverage to provide full-day programs for all needy three- to five-year-olds as well as better teacher training and certification. Private, not-for-profit preschools are normally of high quality, but are expensive and out of reach for many working-class families. Also, even in these schools, despite the expensive tuition, most teachers are underpaid and there is a high degree of teacher turnover.

Many families in the United States must rely on private, for-profit care and early education for their children that is still costly and is often of poor to barely adequate quality, with teachers who lack training and experience. Such programs offer little to kids beyond custodial care. While there have been few ethnographic studies of such child care programs, it is clear that the lack of a planned curriculum, high teacher turnover, and few opportunities to integrate parents and the community into programs work against kids' development of strong and emotionally supportive peer cultures.

The Italian preschools that I studied, although some of the best in the country, are representative of a system with a long history of strong government and civic support. Italian preschools have become even more valued as immigration has increased and children from Africa, the Middle East, and Asia learn Italian and are acculturated through their preschool experience. In addition, in the preschools I studied there is an appreciation and celebration of this growing cultural diversity in Italian society in the curriculum.

Italy, and for that matter all of Western Europe, is different from the United States in many ways. Values in the United States tend to be individualistic while those in Europe are more communal. There is mistrust in the United States of big, federal government social welfare programs that are very common in Europe (for example, universal health care, family leave, child care, and early education programs). Therefore, it is unlikely that a federally supported and administered preschool system like Italy's will develop and be accepted in the United States. However, more and more Americans are becoming convinced of the need for better child care and early education for our children. The increase in funds and bipartisan support for the Head Start program is a reflection of this concern. However, much more investment in Head Start and a raising of the qualifying income level are needed. Some states have begun programs that provide a year of preschool for all four-year-olds (primarily through voucher systems), and many other states are considering such plans. I hope that such plans will provide state subsidies for teacher training and not-for-profit centers as well as regulations regarding teacher training and certification. I also hope that such programs spread to other states, but such diffusion will, no doubt, be slow.

Beyond government programs there are many other things that we can do as individuals and communities to enrich the lives of our children. One thing is to appreciate the complexity, joy, and wonder of kids' peer cultures and the importance of their participation in these cultures for their childhoods. Here we need to provide settings and opportunities for spontaneous free play. Quality early education settings are the top priority for this goal. We also need to be cautious about the overinstitutionalization of childhood that occurs when there are too many structured and formal activities in kids' daily lives. Full-day preschools with plenty of time for free play are needed. But we also need to allow kids to be kids and not overburden them with structured lessons and athletic team involvement, especially during their preschool and elementary school years.

On the other hand, we also need to give kids more of our time and attention. Planned events like visits to amusement parks and family vacations are important parts of childhood. However, what is needed most is everyday time for routine activities, spontaneous play, and talk. Here we need to be more reactive to our kids' interests and let them show us how we can be more spontaneous, curious, and open in our interactions with them.

A major contribution to family isolation among all social class groups in the United States and many other industrialized societies is age segregation. As the anthropologist Enid Schildkrout has noted, the fragmentation of institutions according to age and the high level of social mobility in modern societies "have meant that interaction of persons of different ages occurs less and less frequently and is of diminishing social significance." Most societies will never again experience the close personal relations among generations (in terms of responsibilities and obligations) that prevailed in preindustrial societies. And it is true that many adults do not find a return to such close relations desirable. I'm sure that there are quite a few adults who would find the "kid-free" cruise discussed earlier highly attractive.

Still, older citizens and young children have a lot to offer each other. And in an aging society, the elderly will become more and more dependent on the positive social experiences and socialization of the young. We need to do more to bring children and the elderly together, to share everyday experiences in America. Undoubtedly there are Grandparent Days in schools in the United States as well as in Italy. But these rare occasions need to be expanded to become every-year traditions in all our schools. We also need to reach out to the elderly in our communities, many of whom are isolated in their own homes or retirement centers, to serve as surrogate grandparents for neighborhood children whose own grandparents might live far away. Such programs could involve safe transportation for the children and elderly where public transportation is limited and there are concerns about security. Why should we stop with programs like "Meals on Wheels," when many older Americans need companionship as much if not more than nutrition? We all need more opportunities to engage in routine collective activities with others.

I end with a plea for adults, youth, and kids to break down the barriers of age segregation that exist in modern societies—a plea for us all to take seriously kids' simple request for community: "We're friends, right?"

Notes

· ·

INTRODUCTION

p. 1 **some experts bemoan:** For example, see the following sample of recent books on parenting and childhood, David Elkind, The Hurried Child: Growing Up Too Fast Too Soon. Cambridge, MA: Perseus Publishers. (2001); David Kindlon, Too Much of a Good Thing: Raising Children of Character in an Indulgent Age. New York: Hyperion. (2001); Penelope Leach, Putting Children First: What Our Society Must Do—And Is Not Doing—for Our Children Today; Michael Medved and Diane Medved, Saving Childhood: Protecting Our Children from the National Assault on Innocence. New York: Harper Collins. (1998); and William Sears, Martha Sears, and Elizabeth Pantley, The Successful Child: What Parents Can Do to Help Kids Turn Out Well. New York: Little Brown. (2002).

p. 2 **parents and peers:** Judith Rich Harris, The Nurture Assumption. New York: Free Press. (1998). For a critique of Harris, see D.L. Vandell, Parents, Peer Groups, and Other Socializing Influences. Developmental Psychology, 36(6):699-710 (2000).

p. 3 **Mothers' labor force participation:** U.S. Census Bureau, Statistical Abstract of the United States: (2000); Source for children's attendance of preschool and kindergarten, Digest of Educational Statistics (1997) and Amie Jamieson, Andrea Curry, and Gladys Martinez, School Enrollment in the United States—Social and Economic Characteristics of Students, Current Population Reports, U.S. Census Bureau (2001); Source for number of siblings, Donald Hernandez, America's Children: Resources from Family, Government, and the Economy. New York: Russell Sage Foundation. (1993).

p. 3   **family leave and early education:** I discuss the limits of family leave and child care and early education policies in the United States compared to other modern societies in the final chapter. See Stephanie Coontz, The Way We Never Were: American Families and the Nostalgia Trap. New York: Basic Books. (1992) and Stephanie Coontz, The Way We Really Are: Coming to Terms with America's Changing Families. New York: Basic Books. (1997) for discussions of family change in America.

CHAPTER 1

p. 7   **Betty and Jenny:** Cover names are used for kids, teachers, and schools throughout the book.

p. 8   **Let's start at the beginning:** At the time of my first ethnographic study of preschool children in 1974, there were no published ethnographies of young children in English. Sigurd Beretzen had done such a study earlier in Norway. His study, Children Constructing Their Social World. Bergen, Norway: Bergen Studies in Social Anthropology, No. 36, was published in English in 1984. Since then there have been numerous studies and discussion of methods for studying young children. See Pia Christensen and Alison James, eds., Research With Children: Perspectives and Practices. London: Falmer Press. (2000); and Gary Fine and K. Sandstrom, Knowing Children: Participant Observation with Minors, Newbury Park, CA: Sage. (1988). Some of the material and discussion in this section is adapted from Chapter 1 in Entering the Child's World: Research Strategies for Studying Peer Culture, in my Friendship and Peer Culture in the Early Years. Norwood, NJ: Ablex. (1985).

p. 16 **I was sitting on the floor with two boys (Felice and Roberto):** Some of the material and discussion in this section is drawn from my Transitions in Early Childhood: The Promise of Comparative, Longitudinal Ethnography, in Ethnography and Human Development: Context and Meaning in Social Inquiry. Richard Jessor, Anne Colby, and Richard A. Shweder, eds., 417-457, Chicago: University of Chicago Press. (1996) and William A. Corsaro and Luisa Molinari, Entering and Observing in Children's Worlds: A Reflection on a Longitudinal Ethnography of Early Education in Italy, in Research with Children: Perspectives and Practices. Pia Christensen and Allison James, eds., 179-200. London: Falmer Press. (2000).

p. 23 **The new approaches eschew the individualistic bias of traditional theories:** See my Interpretive Reproduction in Children's Role Play, Childhood, 1(2):64-74 (1993) and The Sociology of Childhood. Thousand Oaks, CA: Pine Forge Press. (1997). See also Alison James, Christopher Jenks, and Alan Prout, Theorizing Childhood. New York: Teachers College Press. (1998) and Jens Qvortrup, Childhood as a Social Phenomenon—An Introduction to a Series of National Reports. Eurosocial Report, No. 36, Vienna, Austria: European Centre for Social Welfare Policy and Research. (1991).

p. 29 **"oppositional talk":** Marjorie H. Goodwin, He-Said-She-Said: Talk as Social Organization Among Black Children. Bloomington: Indiana University Press. (1990).

p. 30 **My first days in the Modena preschool:** Some of the material and discussion in this section is drawn from William A. Corsaro and Luisa Molinari, Entering and Observing.

CHAPTER 2

p. 37   **Much of the traditional work on peer culture has focused on adolescents:** For a classic view of adolescent peer culture from the functionalist perspective, see James Coleman, The Adolescent Society. Glencoe, IL: Free Press. (1961). For more recent studies of preadolescent and adolescent peer culture from the interpretive perspective, see Patricia A. Adler and Peter Adler, Peer Power: Preadolescent Culture and Identity, New Brunswick, NJ: Rutgers University Press. (1997); Donna Eder, School Talk: Gender and Adolescent School Culture. New Brunswick, NJ: Rutgers University Press. (1995); and Paul Willis, Learning to Labour. New York: Columbia University Press. (1981). For discussions of interpretive views of culture more generally, see Clifford Geertz, The Interpretation of Culture. New York: Basic Books. (1973) and also his Local Knowledge: Further Essays in Interpretive Anthropology. New York: Basic Books. (1983). Regarding my definition of peer culture, see my The Sociology of Childhood.

p. 40   **The children's desire to protect interactive space is not selfish:** See my Friendship and Peer Culture.

p. 41   **Here's another example:** See my The Sociology of Childhood.

p. 44   **In Bologna, in the first several months:** See my Routines in the Peer Culture of American and Italian Nursery School Children. Sociology of Education, 61(1):1-14 (1988).

p. 48   **"Climbing bars and other structures ... are designed for children:** For theoretical and empirical work on the body and childhood, see Alan Prout, The Body, Childhood, and Society. New York: St. Martin's Press. (2000).

p. 49   **several children began shouting: "Garbage Man!" "Garbage man!":** The example of "Garbage Man" is drawn from my Friendship and Peer Culture.

p. 52 **"Watch Out for the Monster" Approach-Avoidance Play:** For other discussions of approach-avoidance play, see my *Friendship and Peer Culture* and *Sociology of Childhood*.

p. 57 **In some cases the approach and avoidance phases are repeated several times:** For research on the importance of repetition in discourse and text, see Keenan, Making It Last: Repetition in Children's Discourse. In *Child Discourse*, Susan Ervin-Tripp and Claudia Mitchell-Kernan, eds., 125-138. New York: Academic Press. (1977). For research on prolonging verbal and nonverbal routines, see William A. Corsaro and David Heise, Event Structure Models from Ethnographic Data, *Sociological Methodology*, 20(only one volume per year):1-57 (1990) and Marjorie Harness Goodwin and Charles Goodwin, Children's Arguing. In *Language, Gender and Sex in Comparative Perspective*, Susan U. Phillips, S. Steele, and C. Tanz, eds., 200-248. Cambridge, UK: Cambridge University Press. (1987).

p. 57 **in the Indianapolis Head Start, the kids frequently played a run and chase game they called "Freddy":** For more discussion of the family lives of the Head Start Children, see Katherine Brown Rosier, *Mothering Inner-City Children: The Early School Years*. New Brunswick, NJ: Rutgers University Press. (2000).

p. 62 **Other variants of approach-avoidance have been reported in cross-cultural studies of children's play:** See reviews of children's play and games in Helen Schwartzman, *Transformations: The Anthropology of Children's Play*. New York: Plenum. (1978) and Brian Sutton-Smith's, *The Dialectics of Play*. Schorndoff, Germany: Verlag Hoffman. (1976).

p. 62 *Gaingeen*: See Kathleen Barlow, Play and Learning in a Sepik Society. Paper presented at the Annual Meetings of the American Anthropological Association. Washington, DC (1985).

CHAPTER 3

p. 67   **A big reason that developmental psychologists underestimate the friendship knowledge and skills of young children is:** See William Damon, The Social World of the Child. San Francisco: Jossey-Bass. (1977) and Robert Selman, The Growth of Interpersonal Understanding. New York: Academic Press. (1980).

p. 71   **In these examples ... we have somewhat of a contradiction:** See Thomas A. Rizzo, Friendship Development Among Children in School. Norwood, NJ: Ablex. (1989) for discussion of how conflicts can strengthen close friendships among young children. See also my Discussion, Debate, and Friendship Processes: Peer Discourse in U.S. and Italian Nursery Schools. Sociology of Education 67(1):1-26 (1994).

p. 72   **Friendships, Cliques, and Gender Relations:** For work on gender segregation in children's play, see Hilary Aydt and William A. Corsaro, Differences in Children's Construction of Gender Across Culture: An Interpretive Approach, American Behavioral Scientist. (In press); Ann-Carita Evaldsson and William A. Corsaro, Play and Games in the Peer Cultures of Preschool and Preadolescent Children: An Interpretive Approach, Childhood 5(4):377-402 (1998); Beverly Fagot, Peer Relations and the Development of Competence in Boys and Girls, New Directions for Child Development 65:53-65 (1994); Marjorie Harness Goodwin, The Relevance of Ethnicity, Class, and Gender in Children's Peer Negotiations, In Handbook of Language and Gender. Janet Holmes and Miriam Meyerhoff, eds., London: Blackwell. (In press); Eleanor Maccoby, The Two Sexes: Growing Up Apart, Coming Together, Cambridge, MA: Harvard University Press. (1999); and Barrie Thorne, Gender Play: Girls and Boys in School. New Brunswick, NJ: Rutgers University Press. (1993).

p. 75   **what the sociologist Barrie Throne terms "borderwork":** See Barrie Thorne, Gender Play for examples of borderwork among kindergarten, second, fourth, and fifth grade kids.

p. 75 **"I have a bra for my belly button":** See Hilary Aydt and William A. Corsaro, Gender Across Culture for further discussion of this and similar examples.

p. 79 **"This is the girls' clubhouse!":** See Hilary Aydt and William A. Corsaro, Gender Across Culture.

p. 82 **"Yes, but why do we have to do everything you do, Dante?":** For a more detailed analysis of the friendship dispute between Enzo, Dante, and Mario, see William A. Corsaro and Thomas A. Rizzo, Discussione and Friendship: Socialization Processes in the Peer Culture of Italian Nursery School Children. American Sociological Review, 53(6):879-894 (1988).

p. 83 **a strategy to build solidarity with Mario ... and build a wedge between Dante and Mario:** Attempts to manipulate others in friendship cliques is more common among preadolescent and adolescent children than among preschoolers. See Patricia A. and Peter Adler, Peer Power: Preadolescent Culture and Identity and Marjorie Harness Goodwin, He-Said-She-Said.

p. 85 **In Modena the group of children I joined ... had created a highly communal and rich peer culture:** Some of the material in this section is drawn from William A. Corsaro, Luisa Molinari, Kathryn Hadley, and Heather Sugioka, Keeping and Making Friends in Italian Children's Transition from Preschool to Elementary School. Social Psychology Quarterly, In press.

CHAPTER 4

p. 91 **Almost all definitions of play include:** See Catherine Garvey, Play. Cambridge, MA: Harvard University Press. (1977); L. R. Goldman, Child's Play: Myth, Mimesis and Make-Believe. New York: Berg. (1998); and Helen Schwartzman, Transformations for a discussion of various definitions and theories of children's play.

p. 91 **The expectations kids bring:** Some of the material and analysis in this section is taken from, Jenny Cook-Gumperz and William A. Corsaro, Social-Ecological Constraints on Children's Communicative Strategies, Sociology 11(3):411-434 (1977).

   p. 92   **In spontaneous fantasy:** For other sociolinguistic analyses of children's pretend play, see L.R. Goldman, Child's Play; Marjorie H. Goodwin, He-Said-She-Said; R. Keith Sawyer, Pretend Play as Improvisation. Mahwah, NJ: Lawrence Erlbaum. (1997); and Ursula Schwartz, Young Children's Dyadic Pretend Play: A Communication Analysis of Plot Structure and Plot Generative Strategies. Amsterdam: John Benjamins. (1991).

   p. 94   **The psychologist Jean Piaget characterized:** See Jean Piaget, The Language and Thought of the Child. London: Routledge and Kegan Paul. (1992).

   p. 96   **Danger, Being Lost, and Death-Rebirth:** Material and analysis in this section is taken from my Friendship and Peer Culture.

   p. 100   **these preschool children have recently moved from Piaget's "sensory motor":** See Jean Piaget, The Psychology of Intelligence. London: Routledge and Kegan Paul. (1950).

   p. 107   **The Almost Puppet Show:** Some of the material in this section is taken from my The Sociology of Childhood.

CHAPTER 5

   p. 112   **Child researchers have long argued:** See Dorothy and Jerome Singer, The House of Make-Believe. Cambridge, MA: Harvard University Press. (1990) and Catherine Garvey, Play, for reviews of studies of the positive effects of socio-dramatic role play on children's development. In my work, see especially Friendship and Peer Culture and The Sociology of Childhood, I have argued for an appreciation of role-play as an important element of peer culture. L.R. Goldman, Child's Play; R. Keith Sawyer, Pretend Play; and Helen Schwartzman, Transformations, offer discussions and research similar to my own on children's play. For research on preschool children's role play and gender, see David Fernie, Bronwyn Davies, Rebecca Kantor, and Paula McMurray, Becoming a Person in the Preschool: Creating Integrated Gender, School Culture, and Peer Culture Positionings, in Qualitative Research in Early Education Settings. Amos Hatch, ed. 155-172. Westport, CT: Greenwood. (1995) and Hilary Aydt and William Corsaro, Gender Across Cultures.

p. 113   **In one complex role-play episode from my work in Berkeley:** The material and some of the analysis in this section is drawn from Chapter 3, Children's Conceptions of Cultural Knowledge in Role Play, in my Friendship and Peer Culture and my Children's Conception of Status and Role, Sociology of Education, 52(1):46-59 (1979).

p. 118   **In Piaget's terms:** For a discussion of Piaget's notion of equilibrium, see Jean Piaget, Six Psychological Studies. New York: Vintage. (1968).

p. 118   **As the anthropologist Gregory Bateson argues:** Gregory Bateson, The message "This is play", in his Group Processes: Transactions of the Second Conference. New York: Josiah Macey, Jr. Foundation. (1956). Gregory Bateson as quoted in Helen Schwartzman, Transformations, p.129.

p. 118   **the sociologist Erving Goffman:** Erving Goffman, Frame Analysis. New York: Harper and Row. (1974).

p. 122   **Several ethnographic studies of children's peer culture:** See my Friendship and Peer Culture, pp. 105-120. David Fernie, Rebecca Kantor, and Kim Whaley, Learning from Classroom Ethnographies: Same Places, Different Times, Steven Kane, The Emergence of Peer Culture through Social Pretend Play, In Desire for Society: Children's Knowledge as Social Imagination, Hans G. Furth, ed., 77-97. New York: Plenum Press. (1996).

p. 123   **In my work in Modena, Italy, the five- to six-year old kids:** Some of the material and analysis in this section is drawn from Ann-Carita Evaldsson and William A. Corsaro, Play and Games.

p. 125   **Some child researchers have argued that lower-class children are lacking:** See citations in Schwartzman, Transformations, for a rebuttal of deficit argument. See my Interpretive Reproduction in Children's Role Play.

p. 126   **This is a process I have referred to as "interpretive reproduction":** See my Interpretive Reproduction in Children's Peer Cultures, Social Psychology Quarterly, 58(1):160-177 (1992) and The Sociology of Childhood.

p. 126   **A comparison of the role-play:** Some of the material and analysis in this section is drawn from my Interpretive Reproduction in Children's Role Play. Here I apply the notion of interpretive reproduction to children's awareness of social class differences. For an interpretive view of children's awareness of race, see Debra Van Ausdale and Joe R. Feagin, The First R: How Children Learn Race and Racism. Lanham, MD: Rowman & Littlefield. (2001).

p. 131   **As Peggy Miller and Barbara Moore argue:** Peggy Miller and Barbara Moore, Narrative Conjunctions of Caregiver and Child: A Comparative Perspective on Socialization through Stories Ethos, 17(4):428-449 (1989).

p. 132   **the girls skillfully build coherent discourse through what the anthropologist Marjorie Goodwin terms:** Marjorie H. Goodwin, He-Said-She-Said.

p. 136   **In contrast, Debra and Zena stay very close:** For further research and interpretations of the Head Start materials, see William Corsaro, Luisa Molinari, and Katherine Brown Rosier, Zena and Carlotta: Transition Narratives and Early Education in the United States and Italy, Human Development, 45(5):323-348 (2002). See also Katherine Brown Rosier, Mothering Inner-City Children.

p. 137   **Yet in both cases, these predispositions:** See Pierre Bourdieu, Outline of a Theory of Practice. New York: Cambridge University Press. (1977), for a discussion of the role of predispositions in social reproduction.

CHAPTER 6

p. 140 **what the sociologist Erving Goffman:** See Erving Goffman, Asylums. Garden City, NJ: Anchor. (1961). Some of the material and discussion in this section is drawn from my Friendship and Peer Culture and The Underlife of the Nursery School: Young Children's Social Representations of Adult Rules, in Social Representations and the Development of Knowledge. Gerard Duveen and Barbara Lloyd, eds., 11-26. New York: Cambridge University Press. (1990). See also Amos Hatch, Alone in a Crowd: Analysis of Secondary Adjustments in a Kindergarten, Early Child Development and Care, 44(1):39-49 (1989).

p. 142 **In Bologna during the late afternoon:** See William A. Corsaro and Thomas Rizzo, Disputes in the Peer Culture of American and Italian Nursery School Children, in Conflict Talk. Allen D. Grimshaw, ed., 21-66. New York: Cambridge University Press. (1990) for discussion of preschool children's disputes.

p. 144 **The "*Ci hanno rubato*" routine was:** It was several years later when I presented a talk on these data that an Italian colleague, Paolo Giglioli, pointed out that the correct translation of "*Ci hanno rubato*" is not "They robbed us," but "They stole us" which is, of course, ungrammatical. My colleague thought I knew this, but just translated what the children meant, as I have here. However, this was not the case and I always wondered why Italians laughed so loudly when I presented these data or just described the scene to them. I thought it was funny, but not that funny. The truth is my Italian was no better than that of the children when it came to describing robbing and stealing. In Italian the word for "to steal" is "rubare," and, as is in English, you can steal money, pens, and so on; but you "rob" people, banks, and so on of these goods. In Italian the word for "to rob" is "derubare" which is very similar to "rubare." That's why both the kids and I were confused.

p. 146   **what Goffman calls "working the system":** Goffman, Asylums, p. 210.

p. 146   **That is, the kids used "available artifacts":** Ibid, p. 207.

p. 150   **as Goffman notes, "to work":** Ibid, p. 212.

p. 155   **let's return to the Bolognese kids' "traveling bank":** See William A. Corsaro, Early Education, Children's Lives, and the Transition from Home to School in Italy and the United States, International Journal of Comparative Sociology, 37(1):121-139 (1996) and William A. Corsaro, Interpretive Reproduction in Children's Peer Cultures.

CHAPTER 7

p. 162   **Comparative research of kids' cultures shows:** For research on children's conflict and disputes, see M.P. Baumgartner, War and Peace in Early Childhood, Virginia Review of Sociology 1(1):1-38 (1992); William A. Corsaro, Discussion, Debate, and Friendship; William A. Corsaro, Sociology of Childhood; William A. Corsaro and Thomas A. Rizzo, Disputes in Peer Culture; Marjorie Harness Goodwin, Processes of Dispute Management among Urban Black Children, American Ethnologist 9(1):76-96 (1982); Marjorie Harness Goodwin, He-Said-She-Said; Douglas Maynard, On the Functions of Social Conflict Among Children, American Sociological Review 50(2):207-223 (1985); and Carolyn Shantz, Conflicts Between Children, Child Development 58(2):283-305 (1987).

p. 162   **Among the white middle-class American kids I studied:** Some of the material from this section is taken from Corsaro and Rizzo, Disputes in Peer Culture.

p. 167   **Three five-year-old girls, Ruth, Shirley, and Vickie:** See William A. Corsaro and Douglas Maynard, Format Tying in Discussion and Argumentation Among Italian and American Children in Social Interaction, Social Context, and Language: Essays in Honor of Susan Ervin-Tripp. Dan Isaac Slobin, Julie Gerhardt, Amy Kyratzis, and Jiansheng Guo, eds., 157-174. Mahaw, NJ: Lawrence Erlbaum. (1996).

p. 170 **This type of competitive talk:** See Marjorie Harness Goodwin, He-Said-She-Said and William Labov, Language in the Inner City: Studies in the Black English Vernacular. Philadelphia: University of Pennsylvania Press. (1972).

p. 170 **Here's an example regarding the nature of play:** Some of the material from this section is drawn from my Discussion, Debate, and Friendship.

p. 172 **what the anthropologist, Marjorie Goodwin, calls:** See Marjorie Harness Goodwin, Aggravated Corrections and Disagreement in Children's Conversations, Journal of Pragmatics 7(6):657-677 (1983).

p. 173 **In the Indianapolis Head Start center, several kids:** Some of the material from this section is drawn from my Discussion, Debate, and Friendship.

p. 177 **Not having cable, Alysha remained:** See Katherine Brown Rosier, Mothering Inner-City Children for a detailed discussion of Alysha and her family.

p. 178 **Carlo and Paolo are building:** See William A. Corsaro and Thomas A. Rizzo, Disputes in Peer Culture.

p. 180 **To get a flavor of the complex *discussione*:** For other work on Italian *discussione* among children and adults, see my Sociology of Childhood and Discussion, Debate, and Friendship. See also William A. Corsaro and Thomas A. Rizzo, Discussione and Friendship; Margherita Orsolini and Clotilde Pontecorvo, Children's Talk in Classroom Discussion, Cognition and Instruction 9(1):113-136 (1992) and Clotilde Pontecorvo, Alessandra Fasulo, and Laura Sterponi, Mutual Apprentices: The Making of Parenthood and Childhood in Family Dinner Conversations, Human Development 44(6):340-361 (2001).

p. 180 **Enzo and Mario immediately reject this suggestion:** For a detailed discussion of the debate among Enzo, Dante, and Mario, see

William A. Corsaro and Thomas A. Rizzo, Discussione and Friendship.

p. 182 **In the second example of *discussione*:** For a more detailed discussion of this example, see my Discussion, Debate, and Friendship.

p. 188 **Earlier, I noted that some researchers:** See M.P. Baumgartner, War and Peace in Early Childhood.

p. 189 **The Hair Debate:** See William A. Corsaro and Luisa Molinari, Entering and Observing.

p. 194 **As the anthropologist Roger Abrahams has argued:** Roger Abrahams, Negotiating Respect: Patterns of Presentation among Black Women, in Women and Folklore. Claire R. Farrer, ed., 58-80. Austin, TX: University of Texas Press. (1975), p. 63.

p. 194 **We can develop a better appreciation:** See Roger Abrahams, Positively Black. Englewood Cliffs, NJ: Prentice Hall. (1970); Elijah Anderson, Code of the Street: Decency, Violence, and the Moral Life of the Inner City. New York: Norton. (1999); Marjorie Harness Goodwin, He-Said-She-Said; and Shirley Brice Heath, Ways with Words: Language, Life and Work in Communities and Classrooms. New York: Cambridge University Press. (1983).

CHAPTER 8

p. 202 **However, these programs pale in contrast to:** Medicaid is a health program for economically disadvantaged adults and children. However, a large portion of Medicaid funds is used for full assistance care of the elderly in nursing homes. See Peter Peterson, Gray Dawn. New York: Random House. (1999). Peterson's data are for fiscal year 1995; the disproportion between federal spending for older Americans and children has no doubt increased since then, given inflation in health care costs.

p. 203   **In the United States more than 16 percent:** Child poverty rates and the percentage of children without health insurance are based on reports from the U.S. Census Bureau.

p. 203   **Although the percentage of children living in poverty:** For data on child poverty in industrialized societies, see Timothy M. Smeeding, Lee Rainwater, and Gary Burtless, U.S. Poverty in a Cross-national Context, in Understanding Poverty. Sheldon H. Danziger and Robert H. Haveman, eds., 162-189. New York: The Russell Sage Foundation. (2001).

p. 204   **In countries like France and Italy:** For information on Italy's early childhood education system, see William Corsaro and Francesca Emiliani, Child Care, Early Education, and Children's Peer Culture in Italy, in Child Care in Context: Cross Cultural Perspectives. Michael Lamb, Kathleen Sternberg, Carl-Philip Hwang, and Anders Broberg, eds., 81-115. Hillsdale, NJ: Lawrence Erlbaum. (1992); Carolyn Edwards, Lella Gandini, and George Forman, The Hundred Languages of Children: The Reggio Emilia Approach—Advanced Reflections. Norwood, NJ: Ablex. (1998); and Lella Gandini and Carolyn Pope Edwards, Bambini: The Italian Approach to Infant/Toddler Care. New York: Teachers College Press. (2001). On early childhood education in France, see Barbara R. Bergmann, Saving Our Children from Poverty: What the United States Can Learn from France. New York: Russell Sage Foundation. (1996).

p. 205   **Not-for-profit private preschools:** For studies of the range and quality of child care and early education in the United States, see Cheryl D. Hayes, John L. Palmer, and Martha J. Zaslow, Who Cares for America's Children? Washington, DC: National Academy Press. (1990); Sandra Hofferth, April Brayfield, Sharon Deich, and P. Holcomb, National Child Care Survey. Washington, DC: The Urban Institute. (1991); Sandra Hofferth and Duncan Chaplin, Child Care Quality vs Availability: Do We Have to Trade One for the Other. Washington, DC: The Urban Institute. (1994); and Shankar Vedantam,

Child Aggressiveness Study Cites Day Care, Washington Post, April 19, p. A6.

p. 205   **The Head Start programs I studied:** For further discussion of Head Start curriculum and its effects on the children's transition to elementary school, see William A. Corsaro, Luisa Molinari, and Katherine Brown Rosier, Zena and Carlotta, and Katherine Brown Rosier, Mothering Inner-City Children.

p. 207   **The Italian preschools I studied in Bologna and Modena:** See William Corsaro and Francesca Emiliani, Child Care, Early Education, and Children's Peer Culture in Italy and my Transitions in Early Childhood: The Promise of Comparative, Longitudinal, Ethnography, in Ethnography and Human Development: Context and Meaning in Social Inquiry. Richard Jessor, Anne Colby, and Richard A. Shweder, eds., 417-457. Chicago: University of Chicago Press. (1996).

p. 212   **By civic society I mean:** For work on civic engagement, see Robert D. Putnam, Making Democracy Work: Civic Traditions in Modern Italy. Princeton, NJ: Princeton University Press. (1993) and Robert D. Putnam, The Strange Disappearance of Civic America, The American Prospect, 7(24):344-50 (1996).

p. 214   **While there have been few ethnographic studies:** For an ethnographic study comparing different types of preschool programs in the United States, see Valerie Polakow Suransky, The Erosion of Childhood. Chicago: University of Chicago Press. (1982).

p. 216   **As the anthropologist Enid Schildkrout has noted:** Enid Schildkrout, Age and Gender in Hausa Society: Socio-Economic Roles of Children in Urban Kano, in Sex and Age as Principles of Social Differentiation. Jean S. La Fontaine, ed., 109-137. London: Academic Press. (1975).

Further Reading
· ·

Abrahams, Roger. 1970. Positively Black. Englewood Cliffs, NJ: Prentice Hall.

Adler, Patricia A., and Peter Adler. 1997. Peer Power: Preadolescent Culture and Identity. New Brunswick, NJ: Rutgers University Press.

Anderson, Elijah. 1999. Code of the Street: Decency, Violence, and the Moral Life of the Inner City. New York: Norton.

Coleman, James. 1961. The Adolescent Society. Glencoe, IL: Free Press.

Coontz, Stephanie. 1992. The Way We Never Were: American Families and the Nostalgia Trap. New York: Basic Books.

Coontz, Stephanie. 1997. The Way We Are: Coming to Terms with America's Changing Families. New York: Basic Books.

Corsaro, William A. 1997. The Sociology of Childhood. Thousand Oaks, CA: Pine Forge Press.

Eder, Donna (with Catherine C. Evans and Stephen Parker). 1995. School Talk: Gender and Adolescent School Culture. New Brunswick, NJ: Rutgers University Press.

Elkind, David. 2001. The Hurried Child: Growing Up Too Fast Too Soon. 3rd Edition. Cambridge, MA: Perseus Publishers.

Garvey, Catherine. 1977. Play. Cambridge, MA: Harvard University Press.

Geertz, Clifford. 1973. The Interpretation of Cultures. New York: Basic Books.

Geertz, Clifford. 1983. Local Knowledge: Further Essays in Interpretive Anthropology. New York: Basic Books.

Goffman, Erving. 1961. Asylums. Garden City, NJ: Anchor.

Goffman, Erving. 1974. Frame Analysis. New York: Harper & Row.

Goodwin, Marjorie H. 1990. He-Said-She-Said: Talk as Social Organization Among Black Children. Bloomington: Indiana University Press.

Harris, Judith Rich. 1998. The Nurture Assumption. New York: Free Press.

Kindlon, Daniel. 2001. Too Much of a Good Thing: Raising Children of Character in an Indulgent Age. New York: Hyperion.

Maccoby, Eleanor E. 1999. The Two Sexes: Growing Up Apart, Coming Together. Cambridge, MA: Harvard University Press.

Peterson, Peter. 1999. Gray Dawn. New York: Random House.

Piaget, Jean. 1952. The Language and Thought of the Child. London: Routledge and Kegan Paul.

Prout, Alan, ed. 2000. The Body, Childhood, and Society. New York: St. Martin's Press.

Rosier, Katherine Brown. 2000. Mothering Inner-City Children: The Early School Years. New Brunswick, NJ: Rutgers University Press.

Sears, William, Martha Sears, and Elizabeth Pantley. 2002. The Successful Child: What Parents Can Do to Help Kids Turn Out Well. New York: Little Brown & Co.

Thorne, Barrie. 1993. Gender Play: Girls and Boys in School. New Brunswick, NJ: Rutgers University Press.

Index